THE GREAT CREATE

R RadioShack®

A CELEBRATION OF AMAZING CREATIONS
AND THE PARTS THAT MADE THEM POSSIBLE.

EXTREME LED THROWIES

Build these cheery, magnetic lights — then make them swim, fly, and defy the weather.

The LED "Throwie" was invented around 2005 as a kind of electronic graffiti. It's a little LED light with a strong magnet taped to it. They're fun — you throw them onto metal surfaces and they stick there and stay lit up for days, or even weeks. They're easy, and people love making them.

Since we first published the LED Throwie project, makers have invented dozens of ways to use them. In this project we show you how to adapt Throwies for extreme weather, add a simple "off" switch, and even immerse them in water and float them in the air.

PARTS

For each LED Throwie:
- ☐ LED, 10mm #276-005
- ☐ Coin cell battery, CR2032 type #23-804
- ☐ Electrical tape #64-2373
- ☐ Disc magnet, rare-earth type, 1/2" to 1" diameter

Add for each Indestructible LED Lantern:
- ☐ PVC pipe cap, 1", slip-fit
- ☐ PVC pipe plug, 1", slip-fit
- ☐ Thread sealing tape
- ☐ Binder clip, 3/4" or smaller

Add for each LED Swimmie:
- ☐ Small toy fish, watertight or nearly so

Add for each LED Floatie:
- ☐ Helium balloon, 12", translucent

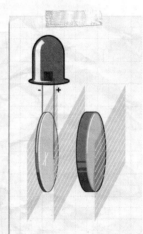

1. Make an LED Throwie. Slip the coin cell battery between the LED's leads so the battery's positive side is touching the positive (longer) lead. The LED will light up. Wrap it once in electrical tape to insulate it. Then stick the magnet on the positive side and wrap it again. You're done. (You can use 5mm LEDs for basic Throwies but you'll want 10mm to make the Indestructible LED Lantern in Step 4.)

Now throw it onto your fridge, or truck, or any ferromagnetic metal surface, and you'll see why they call it a Throwie.

2. Build a Throwie Bug. Throwies naturally stick together magnetically. Chain them together into giant Throwie Bugs to really light things up!

3. Hack your Throwie with an On-Off tab. Cut 2 tabs of cardstock and sandwich the LED's positive lead between them before you tape up the Throwie. One tab will stick to the tape; the other will slip in and out, making and breaking electrical contact. Now you can switch your Throwie on and off.

4. Build an Indestructible LED Lantern. Ditch the magnet and slip your Throwie into a weatherproof capsule made from standard 1" PVC pipe fittings and Teflon tape. These simple, rugged, floating LED lanterns will glow for days, even in extreme weather. They've survived being submerged in water for a week, frozen, and laundered in the washing machine.

5. Make LED Swimmies. Slip your Throwies into little toy fish, seal them up, and set them free in a pond or pool to light up a party.

6. Make LED Floaties. Stuff your Throwies into helium balloons before inflating them. Then cover the ceiling with them, or tether them in bunches anywhere you want cool floating lights.

—*Keith Hammond, MAKE Projects Editor*

To see full build instructions, photos, and video, visit the project page for this build: **radioshackdiy.com/project-gallery/extreme-led-throwies**

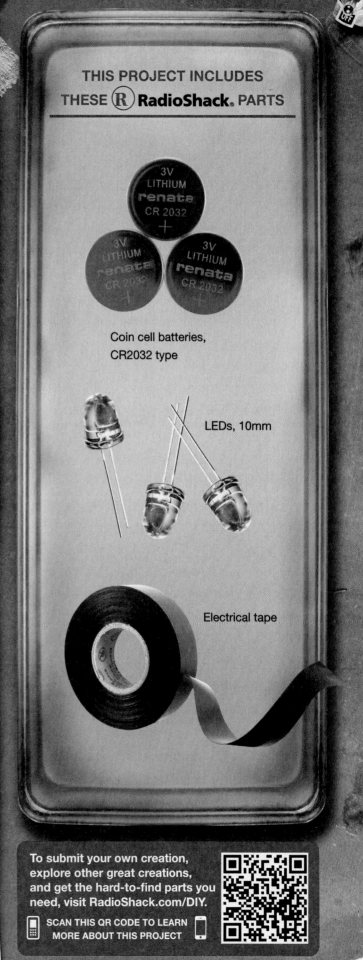

MAKER PROJECTS GUIDE CONTENTS

> "Maker Faire is the only place where someone yells 'fire' and people run *toward* it."
> —Dale Wheat

FOUNDER & PUBLISHER
Dale Dougherty
dale@makezine.com

EDITORIAL DIRECTOR
Gareth Branwyn
gareth@makezine.com

VICE PRESIDENT
Sherry Huss
sherry@makezine.com

EDITORIAL

EDITOR-IN-CHIEF
Mark Frauenfelder
mark@makezine.com

PROJECTS EDITOR
Keith Hammond
khammond@makezine.com

SENIOR EDITOR
Goli Mohammadi
goli@makezine.com

SENIOR EDITOR
Stett Holbrook

TECHNICAL EDITOR
Sean Michael Ragan

ASSISTANT EDITOR
Laura Cochrane

STAFF EDITOR
Arwen O'Reilly Griffith

COPY EDITOR
Laurie Barton

EDITORS AT LARGE
Phillip Torrone
David Pescovitz

DESIGN, PHOTO & VIDEO

CREATIVE DIRECTOR
Jason Babler
jbabler@makezine.com

ART DIRECTION —
SENIOR DESIGNER
Juliann Brown

SENIOR DESIGNER
Katie Wilson

ASSOCIATE PHOTO EDITOR
Gregory Hayes
ghayes@makezine.com

VIDEOGRAPHER
Nat Wilson-Heckathorn

WEBSITE

WEB PRODUCER
Jake Spurlock
jspurlock@makezine.com

MAKER FAIRE

PRODUCER
Louise Glasgow

MARKETING & PR
Bridgette Vanderlaan

PROGRAM DIRECTOR
Sabrina Merlo

SPONSOR RELATIONS COORDINATOR
Miranda Mager

SALES & ADVERTISING

SENIOR SALES MANAGER
Katie Dougherty Kunde
katie@makezine.com

SALES MANAGER
Cecily Benzon
cbenzon@makezine.com

SALES MANAGER
Brigitte Kunde
brigitte@makezine.com

CLIENT SERVICES MANAGER
Sheena Stevens
sheena@makezine.com

SALES & MARKETING COORDINATOR
Gillian BenAry

MARKETING

SENIOR DIRECTOR OF MARKETING
Vickie Welch
vwelch@makezine.com

MARKETING COORDINATOR
Meg Mason

MARKETING ASSISTANT
Courtney Lentz

PUBLISHING & PRODUCT DEVELOPMENT

CONTENT DIRECTOR
Melissa Morgan
melissa@makezine.com

DIRECTOR, RETAIL MARKETING & OPERATIONS
Heather Harmon Cochran
heatherh@makezine.com

BUSINESS MANAGER
Rob DeMartin

OPERATIONS MANAGER
Rob Bullington

PRODUCT DEVELOPMENT ENGINEER
Eric Weinhoffer

MAKER SHED EVANGELIST
Michael Castor

COMMUNITY MANAGER
John Baichtal

EXECUTIVE ASSISTANT
Suzanne Huston

CONTRIBUTING EDITORS

William Gurstelle, Brian Jepson, Charles Platt, Matt Richardson

CONTRIBUTING WRITERS

Martin John Brown, Nicole Catrett, Steve Hoefer, Walter Kitundu, Bob Knetzger, Steve Lodefink, Bre Pettis, Stacey Ransom, Dave Sims

ONLINE CONTRIBUTORS

John Baichtal, Kipp Bradford, Meg Allan Cole, Michael Colombo, Jimmy DiResta, Lish Dorset, Nick Normal, Haley Pierson-Cox, Andrew Salomone, Karen Tanenbaum, Glen Whitney

PUBLISHED BY
MAKER MEDIA, INC.
Dale Dougherty, CEO
Copyright ©2013
Maker Media, Inc.
All rights reserved.
Reproduction without permission is prohibited.
Printed in the USA by Schumann Printers, Inc.

Visit us online:
makezine.com

Comments may be sent to:
editor@makezine.com

CUSTOMER SERVICE
cs@readerservices.makezine.com

Manage your account online, including change of address:
makezine.com/account
866-289-8847 toll-free in U.S. and Canada
818-487-2037,
5 a.m.–5 p.m., PST
Follow us on Twitter:
@make   @makerfaire
@craft   @makershed
On Google+:
google.com/+make
On Facebook: makemagazine

TECHNICAL ADVISORY BOARD

Kipp Bradford, Evil Mad Scientist Laboratories, Limor Fried, Saul Griffith, Bunnie Huang, Tom Igoe, Steve Lodefink, Erica Sadun, Marc de Vinck

INTERNS

Uyen Cao (ecomm.), Eric Chu (engr.), Craig Couden (edit.), Paloma Fautley (engr.), Sam Freeman (engr.), Gunther Kirsch (photo), Brian Melani (engr.), Bill Olson (web), Nick Parks (engr.), Daniel Spangler (engr.), Karlee Tucker (sales/mktg.)

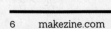

MIX
Paper from responsible sources
FSC® C017648

MAKE CARES
MAKE is printed on 90% recycled, process-chlorine-free, acid-free paper with 25% post-consumer waste, Forest Stewardship Council™ certified, with soy-based inks containing 22%–26% renewable raw materials.

MAKE SPECIAL ISSUE: Maker Faire 2013 is a supplement to MAKE magazine. MAKE (ISSN 1556-2336) is published quarterly by Maker Media, Inc. in the months of January, April, July, and October. Maker Media is located at 1005 Gravenstein Hwy. North, Sebastopol, CA 95472, (707) 827-7000. SUBSCRIPTIONS: Send all subscription requests to MAKE, P.O. Box 17046, North Hollywood, CA 91615-9588 or subscribe online at makezine.com/offer or via phone at (866) 289-8847 (U.S. and Canada); all other countries call (818) 487-2037. Subscriptions are available for $34.95 for 1 year (4 quarterly issues) in the United States; in Canada: $39.95 USD; all other countries: $49.95 USD. Periodicals Postage Paid at Sebastopol, CA, and at additional mailing offices. POSTMASTER: Send address changes to MAKE, P.O. Box 17046, North Hollywood, CA 91615-9588. Canada Post Publications Mail Agreement Number 41129568. CANADA POSTMASTER: Send address changes to: Maker Media, PO Box 456, Niagara Falls, ON L2E 6V2

IT'S ALL ABOUT THE MAKERS

by Sherry Huss, Vice President, Maker Media

SURPRISE. CURIOSITY. FUN. INSPIRATION. MAKE magazine launched in 2005, full of eye-opening how-to projects and fascinating makers. It immediately became the catalyst for a tech-influenced DIY community that has come to be identified as the Maker Movement. Later that year, publisher Dale Dougherty asked, "Wouldn't it be cool if we could get all these makers together in one place to share what they make?"

The result was the first Maker Faire — a gathering of the maker tribe to show and tell, and inspire each other and anyone who'd pay attention. It was an aha moment for us. Today the Maker Movement continues to grow because every day, more people go from observer to participant — inspired by other makers, they begin making things themselves.

We've watched makers gravitate together to form intense creative communities that are innovating in technologies like personal 3D printing, hobby robotics, Arduino microcontrollers, and embedded computing. Makers are showing their work, collaborating online, and egging each other on.

We've seen makers branch outward, launching local makerspaces where any kind of making can be done by anyone — from electronics to CNC wood and metalworking to sewing and other traditional crafts. Makers are sharing expertise and high-tech tools, and cross-pollinating each other's ideas in the process.

And we've watched amateur makers go pro, leveraging shared knowledge and open technologies to manufacture products that can earn them a living. Makers are creating their own market ecosystem of products and services, and learning to push their pet projects out of the nest to fly in global commerce.

A few stories we love to tell:

▶ **SCHOOLTEACHER RICK SCHERTLE** read MAKE, invented an air rocket launcher and wrote a how-to for the magazine, then came and showed it off at Maker Faire. Compressed Air Rockets became an annual Faire favorite for thousands of kids, leading Rick to develop a kit (page 102) — and now thousands more kids have built their own rocket launchers.

▶ **HUSBAND AND WIFE JEFFREY MCGREW AND JILLIAN NORTHRUP** bought one of the first ShopBot CNC routers, taught themselves to use it, and showed off their custom-cut furniture at the first Maker Faire in 2006. Their part-time CNC fascination blossomed into the successful architectural design-build studio called Because We Can. These days when they come to Maker Faire it's to give talks on how to become a professional maker.

▶ **INVENTOR STEVE HOEFER** (page 36) read about the Arduino microcontroller, built an amazing Secret-Knock Gumball Machine and wrote a how-to for MAKE, then came and showed it off at Maker Faire, where thousands of kids (and adults) were delighted by it. He still gets letters from people who were inspired by his project to start making things. (You can build it on page 42.)

▶ **YOUNG ENTREPRENEUR LUKE ISEMAN** (page 34) entered his home-brew electric motorcycle in Austin's Maker Faire and was instantly hooked. Since then he's developed the Garduino garden controller (page 61) and published a how-to in MAKE, leading to his successful Growerbot business. (He launched two other businesses while he was at it.)

▶ **BROOK DRUMM**'s wife bought him MAKE for Christmas, and when he saw a 3D printer on the cover he saved up and bought a kit. Building it with his 6-year-old son, he thought, "I could do better!" — then launched a record-setting Kickstarter campaign. At Maker Faire last year he showed off the portable, ultra-affordable Printrbot Jr. (it won MAKE's "best value" award in our 2012 Ultimate Guide to 3D Printing special issue), inspiring thousands with his product and his story. Thousands more will build Printrbot kits (page 91) and in turn, inspire others with the things they can make.

And that's what it's all about — makers inspiring makers by sharing their methods and projects, whether on the internet, in MAKE, or at Maker Faire, from the San Francisco Bay Area and New York City to Tokyo, Dublin, Singapore, Rome, London, Vancouver, and dozens of other cities around the world. Many are brand-new makers, fired up and ready to build their wildest dream or just solder their first circuit. Others are collaborating online and dreaming of attending a Maker Faire soon to meet up with the maker tribe they're feeling so strongly a part of. Many, like you, are reading MAKE and planning their next project right now.

It's in that spirit that we offer this special edition we're calling the Maker Projects Guide. In it you'll meet some of the makers who've inspired the Maker Movement. You'll build projects they've shared with the world — classics from the pages of MAKE and favorites from Maker Faire. And in our Maker Shed buyer's guide you'll find the kits, books, tools, and boards you need to get started building almost anything you can imagine.

So make something and show it off — online or at a Maker Faire, to your friends, your kids, the world. You never know who you'll inspire.

> "At the heart of Maker Faire is this idea of play. We kind of get lost in it. People here have a love of what they're doing, and it comes across, and you walk away optimistic ... What people come away with is a feeling: they can do things."
>
> —*Dale Dougherty Founder, President, and CEO, Maker Media*

Tom Banwell

FACE TO FACE

Tom Banwell is a self-taught man of many talents. He's a leatherworker, a caster/sculptor, and a tireless inventor of a vast selection of imaginative facemasks, many of which have been featured in films, television, and major magazines.

His most complex and extraordinary works are his "steampunk" gas masks, but he's also known for his delicate, laser-cut leather party masks and other uniquely shaped costume masks. Just to keep things interesting, he also makes rayguns.

His fantastic blog is a must-read for any costume designer or lover of steampunk. It's filled with well-written, step-by-step explanations and interesting tips and tricks. (Be sure to search for "A Steamier Raygun Holster," "Elevated Shoes," and "Modifying a Straw Hat.")

When asked why he gravitated to gas masks, Banwell says, "A gas mask, though functional, dramatically alters the appearance of the wearer. This can be perceived by the viewer as terrifying — as one resembles a monster — or humorous — as one becomes a silly clown."

Banwell manages to combine these two feelings to create unforgettable masks that embody both fear and curiosity. The formal, antiqued leatherwork feels classic and foreboding, but he says the form of the masks — which can resemble a rhinoceros or an elephant — is "pure fantasy."

Banwell is constantly looking at the world around him and re-creating it in the most mad and pleasing manner possible. Looking through his fan photos, it's clear that when seemingly ordinary people don his masks, they unleash the more fantastic selves that lay dormant. —*Stacey Ransom*

tombanwell.blogspot.com

"MAKING is at the center of my teaching."

Jack Chen uses digital fabrication technology to make STEM subjects come to life.

Jack Chen, a former manufacturing engineer and now a Math for America Fellow, is the instructor for the Instrumentation and Automation program, a three-year high school pre-engineering elective at the Sewanhaka (NY) Central School District's Career & Technical Education Center.

Jack says, "A great number of my students are very artistic. They naturally want to make things. What's so exciting about digital fabrication is that you can help students feed their desire to create while learning key STEM concepts."

Jack Chen is one of hundreds of teachers around the country who've joined the 100kSchools.org community, a new, free resource created by ShopBot Tools to help teachers incorporate digital fabrication. "I'm looking forward to finding and sharing projects at 100kSchools, and letting all my teaching colleagues from science to art know about this resource." In this photo: Jack with the ShopBot Desktop he purchased for his classroom as part of ShopBot's "Digital Fab Tools for Schools" promotion that was launched with the support of Autodesk 123D Design.

Jack also serves as the advisor for his high school district's robotics club, the Sewanhaka RoboPandas. In early March, the club competed at the 2013 New York City FIRST Tech Challenge Championship and won! The RoboPandas will now go on to compete at the World Championship in St. Louis in April.

Read more of Jack's story at *www.100kschools.org/blog/* And if you're teaching, whether in traditional schools or other community settings, join us! The 100kSchools community can help you:
- Learn about many digital tools and technologies
- Find projects and curricula and share yours
- Connect with other teachers for mentorship and advice
- Find funding resources for your program

Education Resources for Digital Fabrication

100kSchools is a project of ShopBot Tools, Inc.

888-680-4466 • ShopBotTool

STICK CITY

When **Scott Weaver** first started gluing toothpicks together to create sculptures at the age of 8, little did he know he would later embark on a monumental 34-year journey toward completion of his epic *Rolling Through the Bay* sculpture.

The fourth-generation San Franciscan started *Rolling Through the Bay* in 1974 as a smaller piece that featured his signature ping-pong ball path running through it. He continued to work on the piece off and on until 2008, when he debuted it at the Sonoma County Fair, winning Best of Show. Utilizing a staggering 100,000 toothpicks, it stands 9 feet tall, 7 feet wide, and 30 inches deep, and features four different ping-pong ball routes that start at entry points atop the piece and travel past San Francisco landmarks. Weaver uses only Elmer's white glue.

The ping-pong ball routes are essential for a full appreciation of the details, which are so numerous and uniform in color that they risk being overlooked. The main tour starts at Coit Tower, wraps under a Rice-A-Roni cable car, through the Transamerica Pyramid, out to the Cliff House, down Lombard Street to Chinatown, back toward the Palace of Fine Arts, out around the windmill at Ocean Beach, across the Golden Gate Bridge, over Humphrey the humpback whale, behind Alcatraz, by the Maritime Museum, ending in the long-lost Fleishhacker Pool.

At Maker Faire Bay Area 2011, Weaver earned Editor's Choice blue ribbons and had perhaps one of the most photographed projects at the Faire. He is fueled by seeing people's reactions to his work, recognizing the madness in his method. "What kind of eccentric idiot would spend thousands of hours making a toothpick sculpture? That's me!"

—*Goli Mohammadi*

rollingthroughthebay.com

Luigi Anzivino/Exploratorium

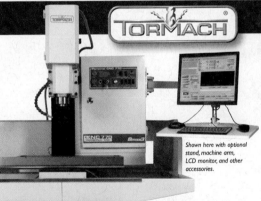

Shown here with optional stand, machine arm, LCD monitor, and other accessories.

Tormach PCNC 770 Series 3 starting at: **$6850**

Tormach PCNC mills are the ultimate maker machines. Whether you're a maker, fabber, innovator, or builder, a Tormach PCNC will enable your ideas with real CNC capability and precision. Don't let your tools hold back your innovation – visit www.tormach.com.

Maker Profile: JDS Labs

Open source design is a key element in the creation of the high-end DIY amplifiers and objective digital-to-analog converters (ODAC) produced at JDS Labs in Glen Carbon, Illinois. Founder and recent Missouri University of Science and Technology graduate John Seaber is using his PCNC 770 to streamline the manufacturing process and produce high-quality product for the image-conscious audiophile marketplace.

"For us, product appearance is just as important as sound quality. Some audiophiles will return a product that performs well because it's visually unpleasant. Cases we machine in-house with the PCNC 770 are of noticeably higher quality than those we had machined by outside shops, because we now have the ability to iterate design and manufacturing changes."

Read the full story at:
www.tormach.com/jdslabs

Tony DeRose

FAST & FURIOUS FUSELAGE

Amid the sea of projects at Maker Faire Bay Area 2012, one shining standout was crafted by a team of five young makers, all under the age of 18 at the time. Welcome to the Viper, a full-motion flight simulator built into the fuselage of a Piper PA-28 plane, complete with 360° rotation on both the pitch and roll axes and a fully immersive flying environment inside. Not your typical after-school project.

Team Viper is **John Boyer**, **Joseph DeRose**, **Sam DeRose**, **Sam Frank**, and **Alex Jacobson**, all members of the Young Makers club. Inspired by a simulator at the National Air and Space Museum, they set out to build a better version based on Battlestar Galactica's Viper spaceship.

Mission accomplished. Once the rider is harnessed in the Recaro racing seat with a full helmet, the plane door is put in place. Inside the cockpit, three 22" high-def screens display the game FlightGear, which you play as you fly. The armrests hold the joystick and thruster, while the custom instrument panel, dozens of buttons and LEDs, and sound system complete the full immersion experience. For control the team used five Arduinos, two iPhones, and one iPad, all networked together. As Sam D. says, "The only senses we don't control are taste and smell — that's for Maker Faire 2013." —*Goli Mohammadi*

the-viper.org

Jason Dietz

WARM GLOW OF ABDUCTION

Inspired by the flying saucers, rocket ships, and robots of 1950s sci-fi comic covers, **Jason Dietz** set out to create a little of that magic for his home. He decided to make lamps that depict a classic flying saucer shooting down a giant plasma ray and pulling up an unsuspecting victim into the ship. To get the desired effect, he knew he had to go big.

Dietz' UFO Lamps stand over 6 feet tall from base to saucer. The 2-foot-diameter flying saucer that crowns each lamp is a sturdy sandwich of parabolic aluminum heat dishes, Edison flame bulbs, and an acrylic disk. The saucer sits atop a giant hand-blown recycled-glass vase that holds 10 gallons of water.

CFL, LED, and halogen lights, in combination with a 110-volt air pump, nail the illusion, as the abduction victim, a lone cow, hovers and twirls helplessly above the grassy pasture from which it was plucked.

With its size, varied lighting, and constant motion, the lamp is beautiful and bizarre at once, not a sight easily overlooked. Dietz keeps one in his living room. "The soft glow of an alien abduction in progress in the corner of the room is quite the sight indeed," he says. "Staring at it for a while lets your imagination run wild — it puts me into that retro sci-fi world."

Like many makers, Dietz gains inspiration as much from seeing his visions come to life as from seeing others enjoying his creations. At Maker Faire Bay Area 2010, he displayed six of his UFO Lamps in a half circle at the back of Fiesta Hall, a dark environment that featured only projects that glow. Fairgoers were drawn in by the UFO beams, and thousands came closer for a good look.

"It looked like a small-scale alien invasion in the back of the hall," Dietz remembers. Apparently he wasn't the only one excited to see this fantasy made reality, as the lamps were in high demand.

"We all have the power to create anything we want to see," proclaims Dietz. "It just depends on how much you really want to see it happen."

—*Goli Mohammadi*

makezine.com/go/dietz

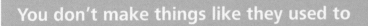

John Edgar Park showcases his Arduino Grande at Maker Faire New York, while co-founder of the Arduino project, Massimo Banzi (background), holds the standard size.

HONEY I SHRUNK THE MAKER

"Arduino is going to be really big at Maker Faire this year." As these words rang in MAKE editor John Edgar Park's ears, the thought occurred to him that he could make Arduino even bigger — literally.

Using 3D modeling software, Park designed a giant version of the Arduino Uno microcontroller board. He then turned his delightful dream into reality by laser cutting, soldering, etching, and painting.

The result? Arduino Grande, a work-ing microcontroller (thanks to a normal Uno mounted on the board) six times larger than life. In the top left corner (the location where a regular-sized Arduino declares its Italian origins), Arduino Grande proudly announces, "Made in Burbank" (Calif.). Park is pleased with his results: "The first time I hoisted it up on my shoulder like a boombox I was pretty darned psyched!"

—Laura Cochrane

makezine.com/go/arduinogrande

Gregory Hayes (top); John Edgar Park (bottom)

Made On Earth
HENNEPIN CRAWLER

If a camel is a horse designed by commit-tee, then what is the Hennepin Crawler?

It looks like a jalopy, but it's really a big bike, designed by **Krank-Boom-Clank**, four Santa Rosa, Calif., artists who wanted to build something that moves as grace-fully along railroad tracks as it maneuvers around the playa at Burning Man.

Two of the members, **Clifford Hill** and **Skye Barnett**, had built an art car for Burning Man in 2007. For the Crawler, they drew in fellow welders **David Farish** and **Dan Kirby**.

"It was a very organic process," says Barnett. "The only thing we had set was that it would be pedal-powered and that it had four seats — since there are four of us."

They also got involved in planning a local event, the Great West End & Railroad Square Handcar Regatta, which aimed to raise awareness for transportation beyond the car, including bikes and com-muter rail. So they designed the Crawler (Farish was fond of the antique-sounding Tom Waits song "9th and Hennepin") to ride the rails, too.

Found materials helped dictate the design: Barnett returned from one dump run with a $15 metal hammock holder. It eventually became the centerpiece of the Crawler's curvy chassis.

"I refer to it as improv, because we were using metal like Play-Doh," says Hill. "We would try something, break it if it didn't work, try something else." They got together once a week — "Our Thursday night TV watching got all screwed up," says Kirby — and cranked into the night to get it finished.

Now they pedal it out to community events, where it draws a lot of interest. "People ask who designed it," says Hill. "Everybody pulled their weight. People can't handle that."

Hill says their goal is to "plug this notion of art and celebration in a public context, inspiring more people to do creative things with bikes, especially kids."

"Kids see it, they find out there's bike parts in it, and then they realize they can make something like that," says Farish.

—Dave Sims

krankboomclank.com

Clifford Hill

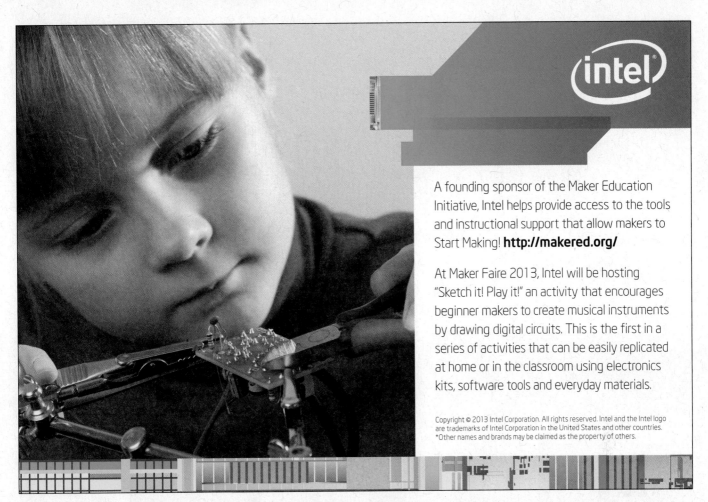

FAIRE *of the* FUTURE

By Goli Mohammadi

> ## "Maker Faire is the mecca of art merged with technology. It's like being at a buffet of knowledge."
> —Justin Gray, artists and roboticist

*N*early 75 years ago, the New York World's Fair of 1939/1940 was the first expo based on the future, with its exhibits geared to give fairgoers a glimpse at "Building the World of Tomorrow." The official pamphlet read, "The eyes of the Fair are on the future ... in the sense of presenting a new and clearer view of today in preparation for tomorrow; a view of the forces and ideas that prevail as well as the machines." Participants included nearly 60 nations and over 1,000 exhibitors, among them the largest U.S. corporations. The next iteration, the World's Fair of 1964/65 continued this glimpse into the future, offering a showcase of mid-century technology and giving many attendees their first interaction with computer equipment, previously tucked away from consumers.

Fast-forward to 2006 and the inaugural Maker Faire, which took place in San Mateo, Calif., featured 100 proud makers, and drew 20,000 folks. Unlike the standard that the World's Fairs set, we turn the power of defining the future to the makers themselves, the creators of the future. We're in our eighth year now, and Maker Faire has grown to become "The Greatest Show (and Tell) on Earth." Our 2012 Bay Area Faire drew 110,000 attendees and featured 900 amazing maker exhibits. There have been countless cutting-edge products launched and tech revolutions, like 3D printing, started at the Faire. The future is here, and it's maker made.

And while our flagship Faires are in the Bay (in May) and in New York City (in September at the original World's Fair site), we encourage community-based, independently produced Mini Maker Faires, both domestic and international, and provide the tools and knowledge to help folks make their own Maker Faires. In 2012, there were 61 Maker Faires, 56 of them community-based Minis and 16 in countries including Japan, Nigeria, Spain, Australia, China, Ireland, Chile, Israel, and the U.K. This fall, Rome is hosting their first Maker Faire, which will span four days and aims to be the biggest Faire in Europe.

Here we take you on a journey into Maker Faires past and present, near and far. Words cannot convey how inspirational and fun attending Maker Faire is. We hope pictures come closer. Find a Faire near you and join us, and if there isn't one, get to know the makers in your community and make a Faire happen. All the information you need is at makerfaire.com.

(Above) Chris James of the R2-D2 Builder's Club at Maker Faire Bay Area 2012. (Below) Elektro the Moto-Man and his dog Sparko, created by Westinghouse for the 1939 World's Fair.

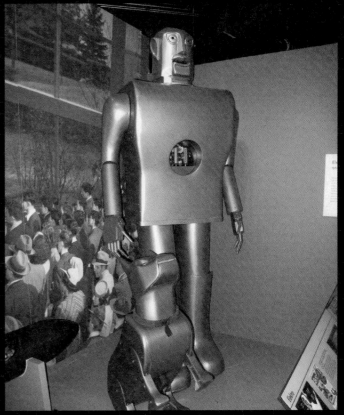

Gary Malerba

Mark Madeo

1. EepyBird's Coke and Mentos fountain show is always a huge crowd pleaser. Here, they create sticky magic at Maker Faire Detroit at the Henry Ford.

2. Artist Robyn Love led an effort to yarnbomb one of the rockets at New York Hall of Science for Maker Faire New York.

3. Todd Barricklow's *Two Penny* is a four-wheeled version of the classic penny-farthing, with 8-foot-diameter wheels and rear-wheel steering. Shown here at Maker Faire Bay Area.

4. Jeff and Lisa Hannan's technicolor *Lunapillar* mobile kinetic sculpture has been giving rides to fairgoers and spreading smiles for years at Maker Faire Bay Area.

5. David Bryan from Minneapolis' Hack Factory takes a spill in their modded vehicle titled "Little Death Trap" at the Power Racing Series race at Maker Faire Detroit.

6. Artist Christian Ristow's 12-foot-tall robotic sculpture *Face Forward* has facial features that are controlled by audience members at remote stations. Shown here at Maker Faire Bay Area.

7. Artist Rob Bell creates giant math-inspired modular Zomes, which are an endless source of amusement for folks young and old at Maker Faire Bay Area.

5

"There's no other
energy like it
on Earth."

—*John Collins, The Paper Airplane Guy*

6

7

Michael C Moore

1. A member of Fakeworks Ladies Racing Team competes at The Madagascar Institute's Chariot Races, which took place at Maker Faire New York 2010.

2. A main component of Maker Faires is the Learn to Solder area, where fairgoers of all ages are taught how to solder, the gateway to electronics.

3. Lisa Pongrace's Acme Muffineering cupcake cars have become an iconic part of Maker Faire Bay Area over the years.

4. Todd Williams' remote controlled, EL wire laden *Land Sharks* have been known to make children squeal as they zoom around Maker Faire Bay Area's dimly lit Fiesta Hall.

5. A fairgoer tries out the glove-like device that controls Christian Ristow's interactive sculpture *Hand of Man*, a 26-foot-long hydraulically actuated human hand and forearm capable of picking up and crushing cars.

6. Artists Ryan Doyle and Teddy Lo's *Gon KiRin* interactive dragon sculpture is built on the frame of a 1963 dump truck. Roughly 64 feet long and 26 feet tall, she spews over 10 feet of flames.

7. Irish roboticist Pete Redmond's RuBot II, seen at Maker Faire New York, can solve a Rubik's Cube in 20 seconds.

8. Five Ton Crane's sci-fi-influenced *Gothic Raygun Rocket Ship* towered 40 feet above the crowd at Maker Faire Bay Area and featured three habitable decks.

9. Lindsay Lawlor's 16-foot *Russell the Electric Giraffe* is a Maker Faire Bay Area favorite and has been to every Bay Area Faire since the first in 2006.

Sabrina Merlo

"Why did we want to participate? How could we not?"
— John Dunivant of Theatre Bizarre

1

2

3

4

> ## "Maker Faire provides a deadline and incentive to build something great before time runs out."
> —*Joseph DeRose, young maker*

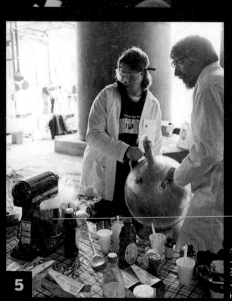

5

6

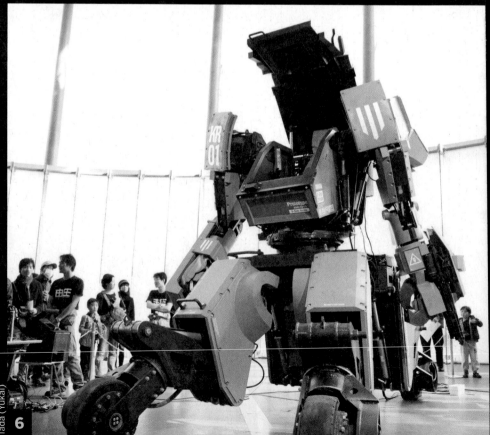

Austin Fresh Photography

Zara Ansar

James Bastow

Tada (Yukai)

7

1. The Austin Bike Zoo displays their art bikes, like this 17-foot butterfly, at Austin Mini Maker Faire in Texas.

2. Artist Ian Langohr displays his larger-than-life masks, like this one, *Nest*, at the Ottawa Mini Maker Faire in Canada.

3. Hands-on activities at the East Bay Mini Maker Faire in Oakland, Calif.

4. The East Bay Mini Maker Faire also features a solid array of art bikes, like this deer bike by Slimm Buick.

5. KwartzLab Makerspace members pour liquid nitrogen from a Dewar into a mixer bowl during a liquid nitrogen ice cream presentation at Mini Maker Faire Toronto.

6. Suidobashi Heavy Industry displays their 13-foot-tall Kinect-operated rideable robot, Kuratas, at Maker Faire Tokyo.

7. A large, stylized anatomical heart sculpture, fresh back from Burning Man, on display at Mini Maker Faire Toronto.

8. A maker displays his Lego domino robot at the inaugural Jerusalem Mini Maker Faire in Israel.

9. Young ones at the Sonoma County Mini Maker Faire in Santa Rosa, Calif., had a blast with the oversized kaleidoscope.

10. The Dublin Mini Maker Faire in Ireland featured this art space shuttle made from discarded tech for an Irish Burning Man decompression party.

Find a Maker Faire near you, or learn how to organize a Maker Faire in your community at makerfaire.com.

8

9

10

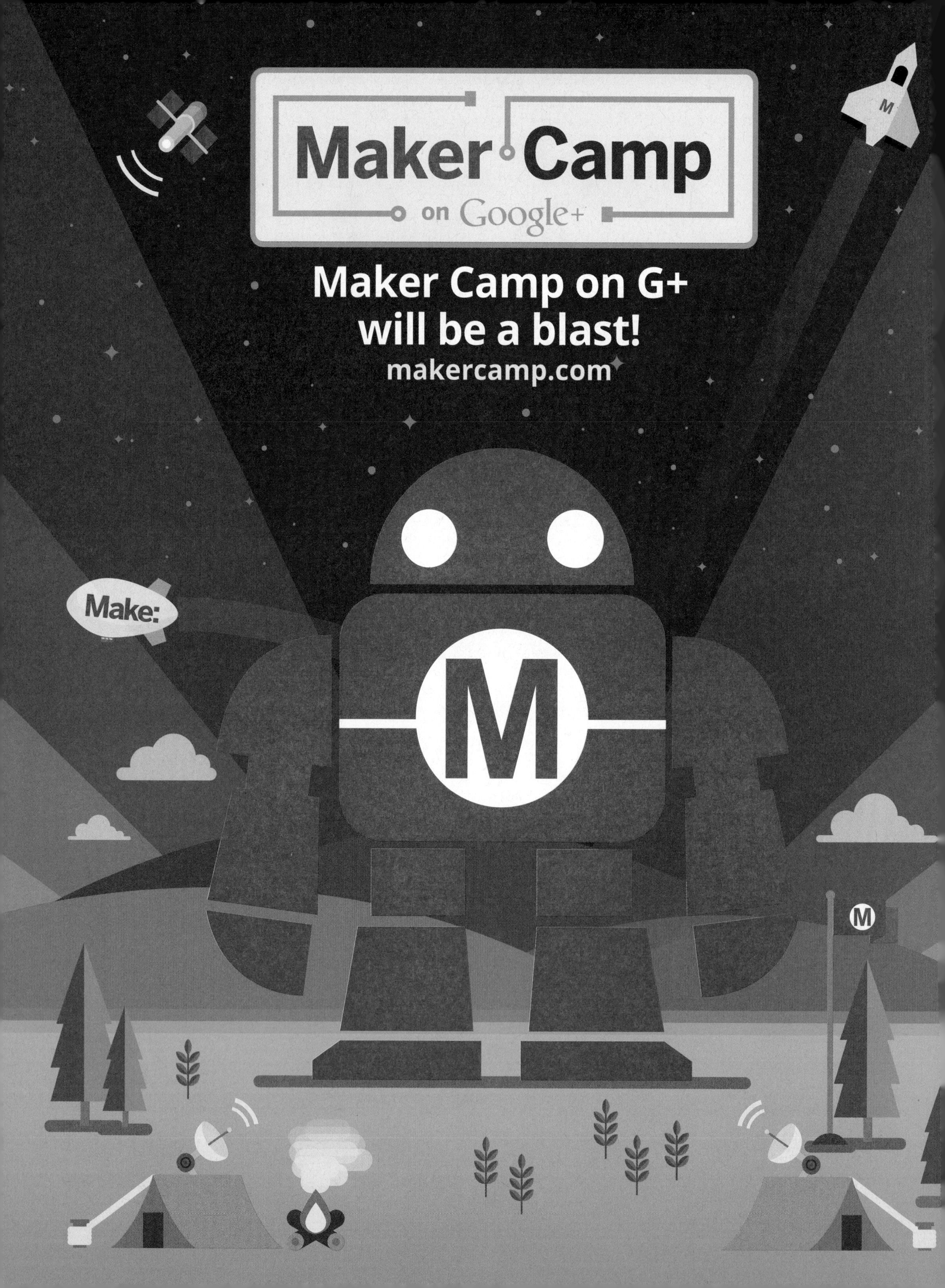

Maker Camp
on Google+
Maker Camp on G+
will be a blast!
makercamp.com
Make:
M

Spend Your Summer Vacation at Maker Camp 2013

Hey campers!

Get ready for 30 days of awesome projects, fun field trips, and your favorite maker celebrities.

No need to bring your sleeping bag or pack your toothbrush, Maker Camp is a free, online summer camp on Google+!

To join, simply create a Google+ profile and follow MAKE on Google+. Or go to makercamp.com and sign up.

Even if you miss a day, you can still experience Maker Camp — just visit makercamp.com for videos of all the cool projects and epic field trips.

JULY

8	WEEK 1 **Makers in Motion**
15	WEEK 2 **Create the Future**
22	WEEK 3 **Fun & Games**
29	WEEK 4 **Art & Design**

AUGUST

| 5 | WEEK 5 **DIY Music** |
| 12 | WEEK 6 **Make: Believe** |

Check out the schedule, and mark your calendars so you don't miss a day! Camp starts Monday, July 8!

ARC ATTACK

By Gregory Hayes

arcattack.com

Gregory Hayes

Andrew Kelly

ArcAttack, an Austin, Texas-based performance group, has made singing Tesla coils famous, appearing at Maker Faires, on television's *America's Got Talent*, and around the world. Their show constantly evolves and improves, thanks to the inspiration and hard work of a rotating cast and crew. We spoke with three of ArcAttack's longest-standing members — Joe DiPrima, brother Giovanni (John) DiPrima, and Steve Ward — to get the story behind the spectacle.

1. HOW DID YOU COME UP WITH THE IDEA FOR MUSICALLY CONTROLLED TESLA COILS?

Steve: I first met Joe in Michigan at a Geek Group meeting, and the idea came up pretty fast. I was playing around with solid state Tesla coil technology, pretty new at the time, and I could control the pitch with a potentiometer. The first time Joe saw this thing, he immediately wanted to control it musically.

2. HOW DOES IT WORK?

Joe: A standard Tesla coil used to have two knobs: pulse rate and pulse width. So I got rid of the interrupter circuit, made a pulse rate modulator out of a keyboard, and played it like a piano.
Steve: The coils we bring to Maker Faire are 14kW outputting 600,000V, and each one makes 10-foot bolts of lightning, which produces the sound. Each time the air is energized it heats up and makes a pressure wave, producing a tick sound. We control the rate, the audio pitch, like 440 snaps a second corresponds to concert A. Anything that can source MIDI will work, whether a computer, keyboard, or MIDI guitar. Joe brewed up some custom hardware, a MIDI player/MP3 player and controller. We compose a MIDI track to be played on the coils while the audio tracks play on the PA system.

3. HOW DO YOU DEVELOP MUSIC TO FIT THE MEDIUM?

John: Well, it's strange. There are a couple of caveats to what we do. No matter what, people are going to like it. *Chicken Dance*? People are going to freak out. But there's not much of a tonality range to use with the coils. You write every song like the lead instrument was a heavily distorted trumpet, and try to make it as pleasing as possible for as long as possible.

I was living in Michigan when Joe called me with this awesome idea, so I'd write the music, sequence it, and send it to him to test it out. I had to compose not knowing how it would work. Like, we didn't have a lot of time when they'd book a show. I'd email it to them while they were on the road; they'd pick it up on wi-fi at McDonald's, test it out, and phone in the changes.

John: These guys are our show brains. I've been trying to convince people the show controller (right) was an iPod prototype from the 80s, but no one believes me. It takes show data from an SD card and outputs from four RCA jacks, and MIDI through a fiber out, and lives next to the sound board far away from the Tesla coils so that EMI from the coils will not be an issue.

We then run a fiber from the show controller all the way back to the stage. With this setup we can get our audio to the sound guy without having any physical connections from the stage to the sound board. The remote (left) links up wirelessly to the show controller via XBee, allowing us to pick songs from the stage without any physical connections. It's totally independent from the show controller so if it screws up around the Tesla coils (it hasn't as of yet), the show controller will still continue to operate.

4. LET'S GO BACK. WHAT GOT YOU STARTED?

Joe: I'm self taught in electronics. Dad, a biomedical technician, taught me a lot and I picked it up hands-on. I always made a habit of taking on projects, using it as a form of education. I was building rudimentary electronics projects by the time I was six or seven, but didn't understand it that well. I took a good interest in electronics until I was 15 or 16, then put four or five years into computer programming. I graduated from high school, needed a job, and got one at a TV repair shop. Then for the next seven or eight years I worked in consumer electronics repairs.

Steve: I had a friend who built a Tesla coil for the 8th grade science fair. I saw it, got hooked on the idea, and started tinkering. I played with high voltage stuff, ways of generating sparks with static electricity. Then in about a year I built my first Tesla coil, slowly pieced together from scrap. I was 13, and got fascinated with making sparks.

John: I always wanted to be a rock star, but never thought it was plausible and studied sound instead.

"There's never anything specific we have

5. WHO ELSE WORKS ON ARCATTACK'S SHOW?

Joe: We've had a few people in and out of our crew and had a lot of people work with us who don't actually travel with us. But let's see. Andrew Mansberger goes on a lot of tours with us, plays guitar and keyboards. Craig Newswanger is pretty much awesome at building anything and built our drum robot. Christian Miller is a computer science guy at UT [University of Texas at Austin], and works with us on a lot of our code. Pat Sullivan's an electrician, goes out with us on shows. He built the Faraday cages.

6. ANY CLOSE CALLS?

Joe: We've never had a dangerous close

call; we're generally pretty safe. We've never almost killed anybody! Oh wait, except this one time. We were at Art Outside in Austin.

Steve: At one show there was a girl, excited to see her friend in the Faraday cage — so excited she hopped our safety fence and ran toward the Tesla coil. Safety engineer Sam McFadden shut the system off; Joe tackled her.

Joe: She just had no idea what was going on, didn't know what she was seeing.

7. WHAT WOULD HAVE HAPPENED?

Joe: We think the arc is probably equivalent to getting hit by a stun gun. I don't think it would kill you instantly.

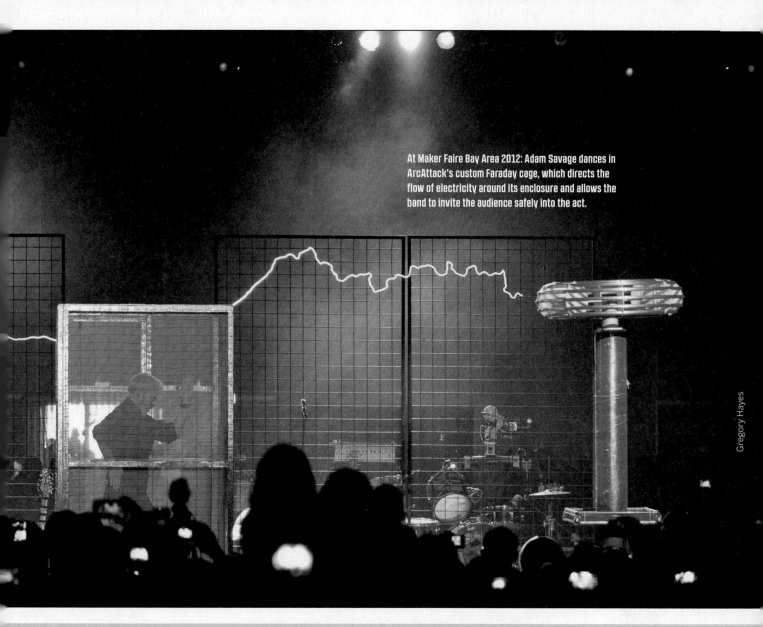

At Maker Faire Bay Area 2012: Adam Savage dances in ArcAttack's custom Faraday cage, which directs the flow of electricity around its enclosure and allows the band to invite the audience safely into the act.

Gregory Hayes

to do ... so we build the things we want to build."

Could put your eyes out or, I don't know, burn you badly. Scare the pants off you. We hope to not find that out.

8. SO WHAT DOES IT FEEL LIKE WITH THE FARADAY SUIT?
Joe: Nothing.

9. NO TINGLE? THE CHAINMAIL DOESN'T GET WARM?
Joe: It feels like nothing at all. If I couldn't see the arcs hitting me, I wouldn't even know it was on.

10. WANT TO ADD ANYTHING — A GOOD STORY?
Joe: We hold a world record for the Tesla coils running during the David Blaine

stunt, a million volts on for 72 hours straight. That's funny; when we first started out, we were just thinking of how to make machines last a week, then two weeks, then machines that would survive shipping.
John: It's extremely difficult to explain what I do in a loud bar. There was an incident where I was trying to describe my seven-foot-tall Tesla coils, but the girl misheard me.
Steve: There is one question I get. If you're really into the Tesla coils, where do you go to learn more? There's no one good place. A Google search will show you lots of stuff. I kind of fumble around on the internet until I find the information. But Tesla himself was

limited to the simple spark gap coil — start with that and work your way toward more advanced stuff.
Joe: I've probably built 20 Tesla coils. Every time we build one we think, "All right, this is it; it doesn't have to be any better." But then, it could be better, so we make it better. ArcAttack is fun because there's never anything specific we have to do. We just have to do something, and it generally works out, so we build the things we want to build.

MEET LUKE ISEMAN

By Stett Holbrook

Boredom is as good an inspiration for making as anything else. Five years ago, Luke Iseman, 30, was working at a startup in Austin, Texas, and he was restless. "I didn't like sitting at a desk. I was very bored."

To break out of his slump, he decided he would rebuild a 1970s Kawasaki to see if he could get it to run on battery power. He picked up a falling-apart motorcycle for under $100 outside Austin and set to work. Never mind that he didn't know what he was doing and had never done anything mechanical beyond changing a flat tire. He liked bikes and thought that was enough to get him started.

He entered the project in Austin's Maker Faire. On the last day of the fair, he got the old bike to work, however briefly. "For a glorious five minutes, I got it to run … I was hooked." He was hooked on making.

lukeiseman.com

Michael T. Carter

"I became more and more excited about not sitting in a cubicle."

Luke Iseman

MAKER MINDED: (Facing) Iseman and his Garduino garden monitor project. (This page) Iseman's electric motorcycle, Re:char's newest biochar kiln design, and the Growerbot, a social, automated garden assistant.

Increasingly, the people he admired were makers, folks who created things and took risks far outside of the office environment where he worked.

"I became more and more excited about not sitting in a cubicle," he says.

Maker Faire nudged him toward a career as a maker, and attending SXSW's interactive panel the following year pushed him over the edge. Two weeks later, he quit his job. Then he took off to Nicaragua, for adventure and to ponder his next move. While roaming around the country, a couple of ideas percolated in his head, and they began to take root once he was back in Austin.

He had previously been exposed to Arduino and loved the idea of physical computing. He was also interested in permaculture and sustainable agriculture. Those interests commingled and yielded Garduino (garduino.dirtnail. com, see also page 61), an open source, Arduino-powered device for monitoring your garden. That project blossomed into Growerbot (growerbot.com), "the world's first social gardening assistant" that helps grow food while offering entertaining updates from the garden.

Meanwhile, his interest in bikes led to another idea: a pedicab business.

While he admits the idea was as much a whim as his trip to Nicaragua, his timing was good, and the business began to build as he refined his designs.

"I had learned just enough to start building them," he says.

Over the course of two years, his fleet of pedicabs grew to 28. He called the company DirtNail, a nickname his former startup boss gave him when he'd come to work with grease-stained fingers during his electric motorcycle tinkering days.

Iseman eventually sold the company to a friend. The fleet of neon orange pedicabs now numbers more than 35. DirtNail was the preferred downtown transportation provider for SXSW 2013.

Before he sold, he'd posted an ad on Craigslist looking for someone to share his underutilized workspace. One of the people who responded was soil scientist Jason Aramburu.

Aramburu had started Re:char (re-char.com), a company that makes biochar (a charcoal fertilizer made from agricultural waste that takes carbon out of the atmosphere) and a simple kiln that allows farmers in developing countries to make it themselves. Iseman, with his background in sustainable ag and his growing maker chops, decided to team up with Aramburu. The two developed a "shop-in-a-box," a portable CNC shop in a shipping container, which they sent to Kenya to train farmers how to make their own biochar kilns. Now Re:char is a bustling business in Kenya.

That company is going strong, and already Iseman is germinating new ideas. Fundamental to his transformation from office worker to maker is the belief that he can make anything.

"It's a material change in perception," he reflects. "For me it's a much more enjoyable and productive way of living my life."

5, 4, 3, 2, 1 THINGS ABOUT
STEVE HOEFER

By Goli Mohammadi

Steve Hoefer is a San Francisco-based inventor and creative problem solver with nearly 20 years of experience. He's contributed several projects to the pages of MAKE, including his Indestructible LED Lanterns, Secret-Knock Gumball Machine, and Haptic Wrist Rangefinder. He's also active in the open source hardware and software communities and is working on a new MAKE video series called Make: Inventions.

grathio.com

Steve Hoefer

Hoefer's simple, rugged, floating Indestructible LED Lanterns project appeared in MAKE Volume 30.

ONE PROJECT YOU'RE PARTICULARLY PROUD OF:

1. The Secret-Knock Gumball Machine. A lot of the things I do are for a specific audience or solving a specific problem, but the Secret-Knock Gumball Machine has something for everyone, and it manages to make candy more fun. It's mechanically and technically pretty simple — you can build your own! I still regularly get messages from people who are inspired by it and have used it as their own springboard into making.

TWO PAST MISTAKES YOU'VE LEARNED THE MOST FROM:

1. The first one is one I *didn't* learn from. My primary and secondary school math teachers were not effective, and I didn't complain when they put me in alternative (non-math) classes. I should've been more involved in my math education and asked my teachers to challenge me more, or simply invested more effort into it. Having a stronger, more confident base in mathematics is something I could use every single day.

2. Second is not bringing in an expert. I've made this mistake more times than I can count and I still fight with it. I come from a very DIY background and I'm really curious, so I want to know how everything works and how to do it. That means that I take on tasks that I hate or are much better suited to a domain expert. I'll spend hours/days/weeks trying to learn how to do something I don't want to do, or that an expert can do better and faster.

THREE IDEAS THAT HAVE EXCITED YOU MOST LATELY:

1. Crowdfunding. Kickstarter gets a lot of love and hate, but the idea is powerful and transformative. It takes the idea of patronage away from popes and kings supporting a handful of artists, and lets anyone support the people and projects that they enjoy. Project flame-outs get a lot of attention, but everyday sites like Kickstarter, Indiegogo, and RocketHub are making things possible that couldn't happen any other way. I'm really keeping my eye on how it will change science.

2. Small-scale or personal factories. Makers creating machines that extend their creativity and power to make. The explosion of hobbyist 3D printers, CNC mills, etc., is creating standardized and affordable motion control. Combining

them with sensors, cameras, increased dexterity, etc., you can create custom machines that make complex objects. A good example is DIWire, an open source CNC-controlled wire bender.

3. Tomorrow. I'm hugely optimistic about the future. The world's not perfect, but historically, earthlings are living longer, healthier, more productively, and more peacefully than they ever have, and I can't wait to see how much further we can go. We're 3D printing cartilage and blood vessels. There is a private space race to mine asteroids. We have flying cars, jetpacks, and hovercraft. My only worry is that reality is outstripping science fiction's ability to make up new things.

FOUR TOOLS YOU CAN'T LIVE WITHOUT:

1. Dremel rotary tool. My Dremel model 395 is durable, portable, and I use it in some way on most projects. I have mounts that turn it into a drill press and a router. I've even used it as a tiny lathe.

2. Ryoba (Japanese pull saw). It has a crosscut blade on one side, rip blade on the other. The blade is thin so the kerf is small, and the blades are replaceable. Since it cuts on the pull of the stroke the saw won't flex when cutting, and it makes straighter cuts with less fatigue than a standard wood saw. My 300mm Gyokucho gets the most use, but I have a 100mm one for special occasions. Great on wood, plastic, and other non-ferrous materials.

3. DSLR. When I want a project to look really nice I can't beat the control and options of a DSLR. At the moment I use a Nikon D5100, which I really like. It's more affordable than the true professional-grade cameras, but has the same sensor and most of the features of the higher-end models. It works with all the past (and future) Nikon lenses and it shoots great-looking video as well.

4. Sleep. I can't count the number of problems that have become manageable or vanished altogether after a good night's sleep. Being well rested gives me patience and optimism I wouldn't have otherwise.

FIVE PEOPLE/THINGS THAT HAVE INSPIRED YOUR WORK:

1. The Apollo program. The brainpower, engineering, research, and design that were applied to the problem were unprecedented. It shows that motivated people can come together to make positive things happen. They solved problems that seemed impossible only a few years before.

2. Benjamin Franklin. Humble, but not afraid to think big. Polymath. Made the best of misfortunes. Constantly worked to improve himself. Most everything he set out to do was for the good of others, yet he rarely lacked for anything.

3. Japan. The current Japanese design aesthetic combines simplicity, functionality, and whimsy, all things I value. (I'm flattered that it goes both ways since my projects have been featured on Japanese TV more often than American TV.) Also, in Japan technology is embraced much more optimistically. Real robots provide opportunities and fictional ones are protectors and companions.

4. The early Royal Society. Hackerspaces have a lot in common with the early Royal Society. Its early members were not professional scientists, but they had ideas, did revolutionary experiments, made devices to prove (or disprove) them, and in doing so changed the world.

5. Kids. They're honest, earnest, curious, and more fearless than adult makers. The stuff they make is incredibly inspirational. And quite often when I see what they create it makes me realize I need to step up my game.

Hoefer's Secret-Knock Gumball Machine, Dizzy Robots, Book Light, and Haptic Wrist Rangefinder.

5, 4, 3, 2, 1 THINGS ABOUT
WILLIAM GURSTELLE

By Goli Mohammadi

Gurstelle's Sound-O-Light Speakers from MAKE Volume 31.

Minnesota-based maker, public speaker, and inventor William Gurstelle has authored seven books, with provocative titles like *Backyard Ballistics*, *Absinthe and Flamethrowers*, and *The Practical Pyromaniac*. He's currently the ballistics and pyrotechnics editor at *Popular Mechanics* and has been a contributing editor to MAKE for years. Gurstelle has taught us how to make many projects, from gravity catapults to flame tubes, and has penned his Remaking History column for the past 3 years, combining history lessons and project builds.

williamgurstelle.com

Gregory Hayes (Sound-O-Light Speakers); Garry McLeod (Flame Tube);
Casimir A. Sienkiewicz (Night Lighter)

ONE PROJECT YOU'RE PARTICULARLY PROUD OF:

1. My favorite project? It's always the one I happen to be working on at the time! Right now, that's a Warren truss bridge. Researching and building this project let me reconnect with the first real engineering course I took in college: Introduction to Statics. Every mechanical, aeronautical, and civil engineer in the world takes that class as a freshman and forgets 90% of it by the time they graduate. For me, designing and building the truss bridge was a great excuse to bone up on a cool but forgotten subject.

THREE IDEAS THAT HAVE EXCITED YOU MOST LATELY:

1. Fast prototyping services are becoming more common. Recently I used a laser-cutting service to get some project parts in a hurry, and they were way better than anything I could cut out of plywood with a jigsaw. Stuff like pro-quality 3D printing, water jetting, and laser cutting is a real boon to makers.

2. Community workshops are wonderful places. Shared tools are great, but shared expertise is even better.

3. Shop class is coming back, sort of. Once something gets thought up, it has

invented if he had access to the camera in an iPhone?

FIVE PEOPLE/THINGS THAT HAVE INSPIRED YOUR WORK:

1. Speaking of Edison, he was quite a maker. Favorite quote: "Just because something doesn't do what you planned it to do doesn't mean it's useless."

2. No one remembers him, but Victor W. Page was the foremost technical writer of the first half of the 20th century. He wrote articles and books on the operation and repair of automobiles, airplanes, motorcycles, tractors, and boats. In

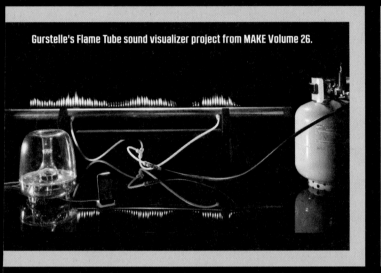

Gurstelle's Flame Tube sound visualizer project from MAKE Volume 26.

The Night Lighter 36 spud gun from Volume 03 was Gurstelle's first project to appear in MAKE.

TWO PAST MISTAKES YOU'VE LEARNED THE MOST FROM:

1. Making bulk buys is often a false economy. Sure, a box of two hundred ¼"x2" machine screws on the internet costs only a little more than buying 25 screws from a hardware store, and buying a gallon of hydrochloric acid doesn't cost much more than a quart. But, my workshop isn't huge and now I've got it packed with stuff that I doubt I'll ever use up. Just buy what you need, when you need it.

2. Using the wrong tool. Yes, I've used a wrench for a hammer, a screwdriver for a chisel, Channellock pliers for a wrench, and so on. Usually it works, sort of, but sometimes it really makes for a bad job. Not real safe either. I've finally learned to go get the right tool for the job, even if I have to walk all the way back to the garage to get it.

to be turned into reality by a carpenter, a millwright, a machinist, a maker. Knowing how to do that is once again cool.

FOUR TOOLS YOU CAN'T LIVE WITHOUT:

1. No tool can do more, ounce for ounce, than a rotary tool like a Dremel. I could probably make a Dremel tool with a few lumps of plastic, steel rod, and a Dremel.

2. I love my hot glue gun — absolutely love it. They say you can have it fast, have it strong, or have it cheap, but you can't have all three. Well, with a hot glue gun, you pretty much can.

3. I also love my hammer. It's an Estwing. While it can't be beat, it sure beats on other stuff.

4. With a digital camera, you no longer have to remember how those parts you're taking apart fit together. A few pics and you've got an instant record. How much faster would Edison have

short, he wrote about anything that was powered by an internal combustion engine. He took what he knew and started an automobile manufacturing company. But he was a much better writer than businessman. See, you can make a living writing about making things!

3. About 20 years ago I saw Christian Ristow and his fiery robots at a show in Phoenix. Right then, I knew I wanted to make stuff like that. His stuff continues to inspire me.

4. It may sound clichéd but it's true: my family is my inspiration. I have a great family — wife, kids, nuclear family, and extended family. That's real luck.

5. It could be self-reverential or at least recursive but MAKE magazine is a tremendous inspiration to me. Every issue contains something that starts my motor running and gets me thinking about making something new.

THE FUN BIKE UNICORN CLUB

By Goli Mohammadi

The Fun Bike Unicorn Club (FBUC), a lively group of bicycle enthusiasts based in Sonoma County, Calif., brought the Death Defying Figure 8 Pedal Car Races to Maker Faire Bay Area in 2012 and 2013. They threw down a challenge to makers: to hand-build a pedal-powered single seater with no less than four wheels, bring it to the Faire, and race it on their tricky track bearing two turns and a cross. The result was a mighty good time had by all. FBUC's focus is on fun and bikes, with a sprinkling of unicorns thrown in for good measure. One of FBUC's founding members, Klaus Rappensperger, gives us the inside scoop.

fbuc.org

John Lohne

UNICORNS UNITE: (Above from left) FBUC's Todd Barricklow and his classic racer at Maker Faire Bay Area 2012, Cyclecide's Laird Rickard joins the fun, FBUC's Joshua Thwaites brings the moxie. (Facing) Klaus Rappensperger rocks the Whiskeydrome and racers work the Death Defying Figure 8 Pedal Car Races at the Faire.

1. HOW WAS FBUC FORMED?

FBUC was formed out of necessity. We (as artists and bicycle junkies) needed a way to display our toys outside of the Handcar Regatta [race in Santa Rosa, Calif.]. We all met racing each other on the railroad tracks and had a great respect for vision, craftsmanship, and what we were doing to make our community more fun. We organized a meeting in my garage one night and wrote down as many random words that epitomized what we do and how we wanted to be portrayed. Then, by random selection (closed eyes and finger points), we selected the four words that make up the Fun Bike Unicorn Club. The idea was that FBUC could stand for anything (Fix Broken Useless Crap — not as cool as unicorns). We claim to be the North Bay Chapter, in hopes that everyone could be a unicorn and start their own chapter. We do not discriminate, nor exclude anyone. Unicorns are everywhere.

2. WHAT TYPES OF EVENTS HAVE YOU PARTICIPATED IN?

We participate in parades, bike expos, nonprofit fundraisers, parties, art openings, galas, events that contact us and pique our interest.

3. TELL US ABOUT THE WHISKEYDROME.

Whiskeydrunk Cycles built the Drome for Joshua Thwaites initially. He found a picture in a book (*The Noblest Invention*) of Keith's Bicycle Track, a sideshow attraction of the late 1800s to early 1900s. We found absolutely no information on the track or who Keith was. This has always haunted me — I gotta know what happened to that track. I feel like

we have such a connection to the four men of the original photo, but know nothing about them. We only found a few photos of it online. I printed out the photos we found and scaled them to the best of my ability. I decided to change the dimensions a bit to make

> "Fun, bicycles, unicorns! Inspire people to put down the remote and pick up a wrench."

it a bit more rideable. Our track diameter is a bit bigger and Keith's is a little steeper. It's not a replica, but more of a homage. We ran into some obstacles and confusion on how they actually built it, so the design was modernized. It breaks down into eight pieces, fits into a 6x10 trailer, and takes about 1-2 hours to set up.

4. WHY A DEATH DEFYING FIGURE 8 RACE?

I saw a small little blurb in a MAKE magazine about cyclekarts [Volume 26, page 54]. These makers were, and are, building little replicas of racecars from the 20s and 30s. They have little 5-horsepower motors in them and they are absolutely beautiful! They race each other and look like they're having a blast. I wanted that. We needed that. The Regatta was done and we needed a new avenue to build, race, and inspire. I put the challenge out to the Unicorns and anyone else who can sit still to build

a car within the requirements. Originally the idea was to build a huge board track and have races, but the figure 8 was easy and way more death defying! For me it was an inexpensive way to fulfill my need to have a vintage racecar. The pedal power just came naturally. We are Junkies of the Bicycle and have a lot of spare parts.

5. HOW MANY MEMBERS DOES FBUC CURRENTLY HAVE AND HOW OFTEN DO YOU MEET?

Whiskeydrunk Cycles meets on Wednesday nights. The Unicorns crash the meetings on occasion. Currently there are five organizations that make up the North Bay Chapter: Klank-Boom-Crank, T3D, Eight Pack, Bunnyfluffer Cycles, and Whiskeydrunk Cycles.

6. WHAT IS THE FBUC PHILOSOPHY?

Fun, bicycles, unicorns! Inspire people to put down the remote and pick up a wrench.

7. HOW MANY YEARS HAVE YOU BEEN MAKERS AT MAKER FAIRE AND WHAT KEEPS YOU COMING BACK?

Four years. It's a great platform for inspiration! It boggles my mind to see the amount of people who walk through those gates and see stuff that they would have never found if it were not for MAKE magazine. The look of awe on the children's faces is what keeps me motivated to continue doing what I do. I had *Popular Science* and the Boy Scouts when I was a kid, but nothing even close to what MAKE has done. I still haven't seen everything the Faire has to offer.

Mike Solari

Garry McLeod

SECRET-KNOCK GUMBALL MACHINE

By Steve Hoefer

CANDY RAPPER—Make a cute candy vending machine that only dispenses treats when you knock the secret rhythm on its front panel.

One of the best things about exhibiting at Maker Faire is giving attendees a challenge. For the 2010 Maker Faire Bay Area, I decided to combine a past project of mine, a door lock that opens only when you give a secret knock, with a standard crowd pleaser: candy.

The result was this Secret-Knock Gumball Machine, which tempted and tested the crowds at Maker Faire to guess the right rhythm and receive a treat. Since the knock was not terribly secret (I happily handed out hints), it distributed hundreds of gumballs over the event's two days.

The "secret" knock defaults to the famous "Shave and a Haircut" rhythm, but you can program custom knocks by simply pressing a button and knocking a new pattern. The machine only listens for the rhythm, not the tempo, so the correct knock will dispense a treat whether you perform it fast or slow.

Inside the machine, a piezo sensor picks up sounds from the front knock panel, while an Arduino microcontroller recognizes the target pattern and controls a servo-driven gumball-dispensing wheel. You can build the Secret-Knock Gumball Machine with its inner workings visible or hidden, depending on whether you want to show off the mechanism or keep it a mystery.

OPEN SEZ ME

A Sweet, satisfying gumballs tempt passersby.

B A piezoelectric sensor detects knocks on the front panel.

C An Arduino microcontroller listens for the knocks and tracks the relative times between them.

D If the Arduino determines that the input knock rhythm matches the target pattern, it switches on the servomotor.

E The servo rotates the dispensing wheel through the gumballs, which causes one of them to fall into its delivery hole.

F As the wheel completes its rotation, it dumps the gumball from its delivery hole into the delivery tray.

G Whenever the Arduino hears a knock, it also flashes the green indicator light.

H You can program the Arduino to listen for different patterns by pressing the programming button on the back and knocking a new sequence.

I The Arduino shines the red indicator light while the programming button is pressed and whenever a knock pattern fails to match the target.

J The sensitivity potentiometer sets the level of loudness the circuit will recognize as a knock.

K All project-specific circuitry is built onto a perf board shield that plugs on top of the Arduino. This lets you unplug the microcontroller for use in other projects.

BONUS SPRAY PAINT STENCILS

Download stencils of this design by Rob Nance at makezine.com/25/gumball.

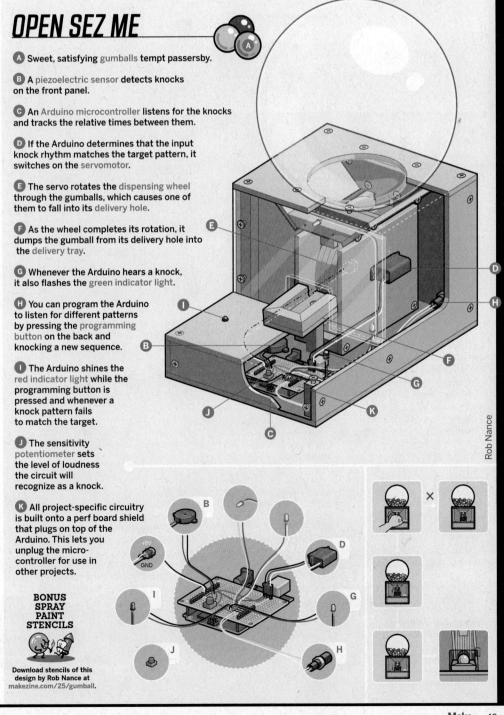

Rob Nance

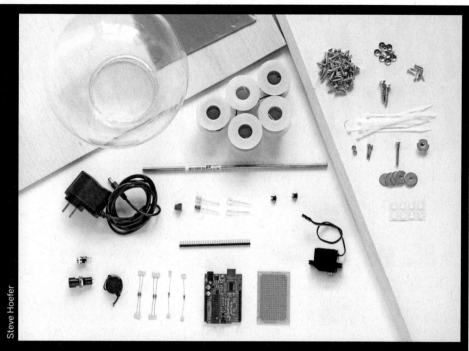

Steve Hoefer

MATERIALS

» **Plywood, ¼" thick, or 6" clear acrylic, 2'×4'** Use plywood to hide the internal workings or acrylic for maximum visibility. Plywood is much easier to work with than acrylic, which tends to crack, chip, and scuff. To give a view inside with minimal hassle, I cut all my pieces out of plywood except for the 7¾"×5¹⁄₁₆" acrylic access panel in front.

» **Wood stock, ½" square, 64" long**

» **Clear plastic light globe, 8" diameter with 4" opening** item #3202-08020 from 1000bulbs.com, $11

» **Wood screws: #8, ¾" long (65); and #12, 1¼" long (2)**

» **T-nut, #10-24, long prong**

» **Machine screw, #10-24, 1¼" long**

» **Washers, ¾" OD, 6" ID (5)**

» **Brass tube, ⁷⁄₃₂" diameter, at least 2" long**

» **Arduino Uno microcontroller** item #MKSP11 from Maker Shed (makershed.com), $35. You can also use an older Arduino Duemilanove or equivalent, but in one test build an Uno worked while the Duemilanove mysteriously did not.

» **Servomotor, small, high-torque** Turnigy S3317M, #TGY-S3317M from hobbyking.com, $8

» **Male breakaway headers, 40×1 pin**

#PRT-00116 from SparkFun Electronics (sparkfun.com), $2

» **Perf board, 0.10" pitch, at least 15×20 holes** #276-149 from RadioShack (radioshack.com), $3 (or cut from a larger piece)

» **LEDs, 5.5mm: red (1), green (1)**

» **LED holders, panel mount (2)** RadioShack #276-079, $2 each

» **Piezo buzzer, 1" diameter** RadioShack #273-059, $4

» **Resistors, ¼W: 100Ω (2), 150Ω (2), 10kΩ (1), and 470kΩ (1)**

» **SPST momentary push-button, panel mount** RadioShack #275-618, $3

» **Potentiometer, 10kΩ linear trim** SparkFun #COM-09806, $1

» **Power jack, 5mm×2.1mm coaxial (size M), panel mount** RadioShack #274-1582, $4

» **9V DC power supply with 5mm×2.1mm center-positive plug** Maker Shed #MKSF3, $7

» **Insulated solid-core wire, 22 gauge, multiple colors, around 8' total length** plus another 5' if making the transparent case

» **Stick-on rubber feet (8)**

» **Small cable ties (8)**

» **Wood screws, round head, #2, ½" long (6)**

» **Machine screws, round head, #4-40, ½" long (2)**

» **Nuts, #4-40 (2)**

» **Gumballs, 1" diameter** a standard box of 850

[OPTIONAL, FOR VISIBLE INTERIOR]

» **LEDs, high brightness, white (2)**

» **Thumbtacks (2)**

TOOLS

» **Drill or drill press and assorted drill bits**

» **4" hole saw**

» **1" Forstner (flat bottom) drill bit**

» **Wood saw(s)** for cutting straight lines and curves; a table saw and jigsaw, or a handsaw and miter box

» **Vise**

» **Screwdrivers**

» **Hammer**

» **L-square ruler**

» **Pencil**

» **Masking tape**

» **Pipe cutter or hacksaw**

» **Sandpaper, 80- and 120-grit**

» **Wire cutters and strippers**

» **Needlenose pliers**

» **Thick cardstock or cardboard, 6"–12" square**

» **Soldering iron and solder**

» **Computer with printer and internet connection**

» **USB cable**

BUILD YOUR SECRET-KNOCK GUMBALL MACHINE

1. BUILD THE CASE

1a. Download the templates from makezine.com/25/gumball and print them at full size. Transfer to plywood or acrylic, and cut all pieces to size. Mark all screw and rail locations for later drilling and placement. To avoid confusing or flipping parts, temporarily mark the outside of each part with masking tape. Sand all edges smooth

Use a 4" hole saw to cut the hole in the top plate as well as the 7 dispensing wheel disks (2 stir plates and 5 center disks). You can use the remnant of the top plate hole as one of the dispensing wheel disks. Drill the ¼" axle holes (which are off-center on the stir plates) as shown on the template.

> **TIP:** To minimize chipping when working with acrylic, support the work well, tape both sides of all cuts, and use fine-bladed saws at low speed — or have your local plastics retailer cut the pieces.

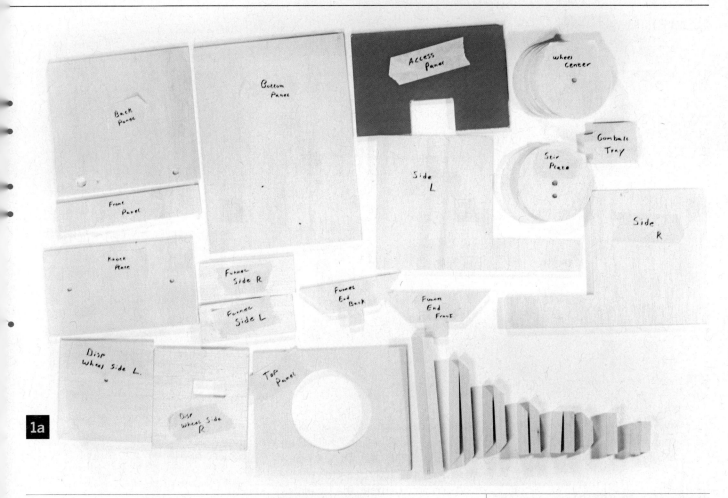

1a

1b

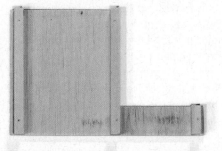

1c

1b. Attach 3 rails to each side panel where indicated on the template. The front and back rails are ³⁄₁₆" from the edge. The 2 side pieces should mirror each other, with rails on the inside of the case.

NOTE: Unless otherwise specified, all attachments in this project are made with #8×¾" wood screws, and pilot holes should be drilled to prevent splitting.

1c. Attach the bottom panel and front panel to the side panels.

1d. Drill holes for the power plug and programming switch in the back panel, then attach it to the side panels.

1d

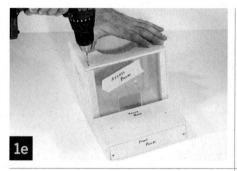

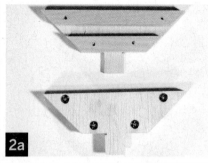

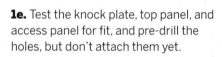

1e. Test the knock plate, top panel, and access panel for fit, and pre-drill the holes, but don't attach them yet.

2. BUILD THE GUMBALL FUNNEL

2a. Attach the top and bottom rails to each funnel end piece.

2b. Attach the funnel side pieces to the funnel ends. Be sure to put the rails on the outside of the funnel.

2c. Attach the completed funnel to the bottom of the top plate with 4 screws.

3. BUILD THE DISPENSING WHEEL

3a. Make a sandwich of round plates: the 5 dispensing wheel plates stacked in the middle and one stir plate on each end, with the stir plates rotated 30° from each other. Push the brass tube through the axle holes and use an L-square to check for true.

Clamp, pre-drill, and secure the whole stack with a #12×1¼" wood screw on each side.

3b. Clamp the dispensing wheel (in a drill press if available) and drill a 1" deep hole using a 1" Forstner bit. The hole should point directly down toward the axle, centered between the stir plates and equidistant between their peaks. To help gauge the right hole depth, place a masking tape flag 1" from the bottom of the drill bit.

Toss a 1" gumball in the hole to check for fit. It should fall in and out easily, and the top of the gumball should sit flush with the wheel.

3c. Attach the servo horn (the X-shaped plastic piece that connects to the servo shaft) to the center of the wheel using four #2×½" screws. Space the screws so they don't intrude into the 1" hole drilled in the previous step. To center the horn, sight its drive-shaft mounting hole through the axle hole of the dispensing wheel.

4. MOUNT THE DISPENSING WHEEL

4a. Before attaching the servo, we need to zero its rotational position using the Arduino. Download and install the latest version of the Arduino

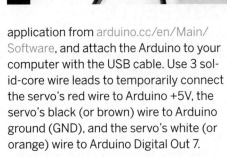

4a

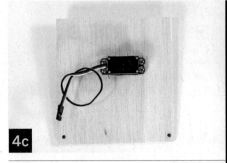

4c

4d

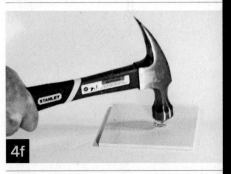

4f

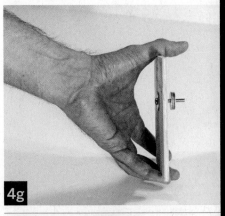

4g

4h

application from arduino.cc/en/Main/ Software, and attach the Arduino to your computer with the USB cable. Use 3 solid-core wire leads to temporarily connect the servo's red wire to Arduino +5V, the servo's black (or brown) wire to Arduino ground (GND), and the servo's white (or orange) wire to Arduino Digital Out 7.

4b. Download *servo_reset.pde* from makezine.com/25/gumball, open it in the Arduino application, and upload it to the microcontroller. The servo should rotate fully one way, then the other, then stop.

4c. Remove the wires connecting the servo to the Arduino. Use the servo's supplied mounting screws to attach it to the dispensing wheel right side plate, with the servo shaft centered to line up with the 7/32" center hole in the left side plate. Before screwing, fit the included rubber bushings and metal grommets into the servo mounting holes to provide strain relief, with the curved edges of the grommets facing the wood. It's OK if the screw points run out the other side of the plate.

4d. Orient the dispensing wheel next to the right side plate so that the gumball hole tilts back 45° toward what will be the back of the machine. Slide the servo horn over the servo shaft and secure it with the included servo horn screw that runs through the center of the horn.

4e. Use a hacksaw or pipe cutter to cut a 1 7/16" length of brass tube for the axle. Insert the tube into the dispensing wheel axle hole. It should be flush with the stir plates.

4f. Hammer the T-nut firmly into the hole on the dispensing wheel left side plate, on the outside (not the wheel side).

4g. Place a washer on the #10-24x 1¼" machine screw, thread it entirely through the T-nut, and tighten the bolt firmly. Place three 3/16"×¾" washers on the inside of the bolt.

4h. Insert the end of the bolt into the dispensing wheel axle. Attach the left and right side plates to each other with the 2" dispensing wheel rails. If there's a lot of play between the T-nut and the dispensing wheel, add an additional washer before screwing the plates together.

5. ASSEMBLE AND TEST THE ELECTRONICS

For all connections, refer to the schematic and circuit board layout at makezine.com/25/gumball.

5a. Cut and strip leads for all off-board components. You'll need four 5" leads for the indicator LEDs and four 14" leads for the power connector and programming

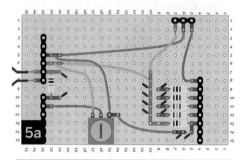

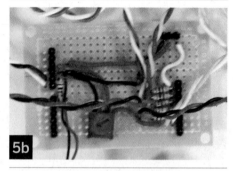

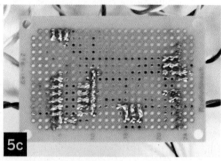

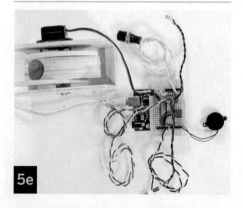

button. If you're making a transparent case, you'll also need two 12" leads and two 16" leads for the white LEDs.

5b. Cut the breakaway header pins into rows of 8, 6, 6, and 3. On all but the 3-pin header, use needlenose pliers to push the pins flush with the plastic spacer.

5c. Solder the headers in place on the perf board, following the layout diagram *SNGM_layout.pdf*. The 3-pin header's long pins should point up, for connecting the servo, while the other header pins should run down through the bottom of the board for plugging into the Arduino.

5d. Solder the rest of the components to the perf board following the layout or schematic diagram *SNGM_schematic. pdf*. Pay special attention to the polarity of the LEDs (longer leg is positive) and power connector (center pin is positive). Omit the 2 white LEDs and the 100Ω resistors if you're not making a transparent case.

5e. Plug the completed shield into the top of the Arduino, aligning the 8-pin header with digital pins 0–7. Plug the servo into the 3-pin header on the shield, taking care to observe the polarity (because plugging the servo in backward can damage it).

5f. Connect the Arduino to your computer with a USB cable. Download the code file *secret_knock_gumball_machine. pde* from makezine.com/25/gumball and upload it to the Arduino.

After you upload the sketch (or apply power for the first time) all the LEDs should light, and the servo should rotate the dispensing wheel to its start location. When the lights go out, it's ready to listen for a knock.

5g. Turn the sensitivity potentiometer down until the green LED stops blinking. If the light stays on no matter how far you turn it, check the piezo sensor wiring for shorts or bad connections.

5h. For testing, adjust the pot so that the green light stops blinking without any

input, but gently tapping the knock sensor makes it blink.

Tap the "Shave and a Haircut" rhythm. The green LED should blink and the servo should rotate. If you included the white LEDs, #1 should blink and #2 should stay lit for a few seconds. If you entered an incorrect rhythm, the red LED should blink.

Press the programming button, and both red and green LEDs should light. Tap a new rhythm, release the button, and wait a second. The red and green lights should echo the rhythm visually, and tapping this new rhythm should rotate the servo. Resetting or unplugging the Arduino will revert it to the default "Shave and a Haircut."

6. WIRE AND COMPLETE THE CASE

6a. Secure the top panel to the rest of the case with 4 screws. Align the panel so the offset 4" hole is toward the back of the device.

6b. Secure the dispensing wheel assembly to the bottom panel with 2 screws. Rotate the dispensing wheel to check for clearance against the back panel and the funnel, and remove material wherever it hits the wheel.

6c. Attach the gumball tray support directly in front of the dispensing wheel, with the support's angled end on top, and inclining down toward the front of the case.

6d. Unplug the shield from the Arduino. Attach 4 rubber feet to its underside and use two 4-40 bolts and nuts to attach it to the bottom panel. Insert the screws up from the bottom through the pre-drilled holes, and thread the nuts on top. Do not overtighten the nuts.

6e. Attach the knock sensor to the center back of the knock plate using two #2×½" screws. The piezo buzzer should fully contact the back of the knock plate, and the screws shouldn't penetrate through to the other side of the wood.

6a

6b

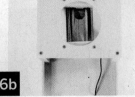

6c

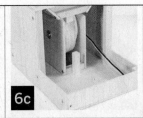

6d

6e

6g

6h

6k

I attached the sensor "upside down" with its mounting tabs away from the wood to keep the screws from running through, but the sensor will function facing either way.

6f. Insert the LED holders into the front of the knock plate and slide the red and green LEDs into the holders from the back.

6g. Attach the power plug and programming button to the back of the case with their mounting hardware.

6h. If you're making a transparent case, attach the white LEDs to illuminate the action. Tighten a small cable tie around the base of each LED and secure it in place with a thumbtack. Place LED #1 behind the right side pillar pointing inward, and LED #2 on the bottom front of the dispensing funnel, pointing to the gumball tray. Use cable ties to secure and manage the cables inside the case.

6i. Screw the knock panel into place.

6j. Screw the gumball tray sides to the gumball tray, and screw the tray down onto the tray support post. Before attaching it permanently, check for gumball clearance against the dispensing wheel and the access panel.

6k. Stick a rubber foot on each bottom corner to prevent the machine from sliding around as people knock. Finally, attach the access panel. You're done!

Visit makezine.com/25/gumball **for project templates, Arduino code, a schematic diagram, and a layout/ wiring diagram.**

KNOCK YOURSELF OUT!

MACHINE SETUP

1. Plug the power adapter into a handy outlet and plug the other end into the back of the machine.

2. Fill the plastic globe with your favorite 1" gumballs (or other candy balls) and put it on top. To prevent spilling the balls all over the floor, put a piece of thick cardstock or cardboard over the hole in the globe, invert it in place, then slide the card out.

3. Program your new secret knock or leave it with the standard "Shave and a Haircut" knock.

4. Knock and enjoy!

> **TIP:** To make the machine more responsive, loosen the screws that hold down the knock plate and gumball tray, so they move and clatter when you knock.

TREAT SELECTION

The 1" ball is a vending industry standard for not only gum, but also jawbreakers and other candies, bouncy balls, toy capsules, and empty capsules that you can fill with your own small objects.

An 8" globe holds about 200 gumballs, which are usually sold in quantities of 850 — but you can also buy 1" gumballs by the pound at bulk candy stores, and there are sellers on eBay that offer them in smaller quantities. I also found a box of 500 gumballs at Smart & Final. ◪

Steve Hoefer is a technological problem solver in San Francisco. He spends much of his time trying to help people and technology understand each other better.

Liz Smith (bottom)

SKILL BUILDERS:
SOLDERING AND DESOLDERING

Soldering is a way of joining metals using soft alloys that melt at lower temperatures. It's especially useful in electronics for making permanent, electricity-conducting connections between components. Desoldering is just as handy, especially for salvaging parts from old boards. Both skills are easy to pick up with a few pointers.

ETCHING PRINTED CIRCUIT BOARDS

Printed circuit boards, or PCBs, are the foundation of most modern electronic devices. Using a laser printer, an electric iron, and a few other simple tools, it's relatively easy to make small batches of custom PCBs at home, and give your projects a professional look that's prettier than perfboard by a mile.

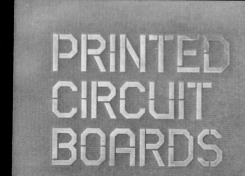

Indestructible LED Lantern

LED Throwies, a kind of electronic graffiti, are a favorite project at Maker Faire. Build them from just an LED, battery, magnet, and tape, then throw them and they'll stick to any ferromagnetic surface. Fun! Now take them up a notch: make an Indestructible LED Lantern to defy any weather. (See page 2 for more Extreme LED Throwies ideas.)

One-Hour CoasterBot

Build a basic programmable robot platform in about an hour using a tiny Arduino clone, some hacked servomotors, and a couple of dead CDs (aka "coasters"). Then start experimenting! Add new sensors, actuators, and other features as you expand on the basic bot-building knowledge you learn in this project.

180

SKILL BUILDER:
SERVO CONTROLLERS

Servomotors do the heavy lifting in countless projects. Because the motors themselves are so visible, it's easy to overlook their controller behind the scenes. While choosing servos is fairly simple, choosing a controller is bit more complicated. How many servos can you run at once? Do you need a dedicated controller? Read our guide and find out.

The Brain Machine

Another Faire favorite, the Brain Machine commandeers your mind with lights and sounds that pulse at brainwave frequencies: alpha, beta, theta, and delta. Build the kit from the Maker Shed, then train your brain for meditation, relaxation, even hallucination. To hack the kit for a stealthier look, mount the circuit in a mint tin instead of on the glasses.

The Six Pack Tesla Coil

This classic spark-gap Tesla coil is smaller than ArcAttack's design, but still packs a wallop, and will draw a 15" spark through open air. It uses an old neon sign transformer and a capacitor bank made from foil-wrapped beer bottles. If you're old enough to buy the beer, you're old enough to build the coil!

Kitty Twitty

Build a networked cat toy that tweets when touched. Even if you don't have a kitty, this technology is easy to adapt for other sensing jobs. Want to know when the garage door opens? When there's water on the basement floor? If you've got internet access at both ends, this project makes it easy.

Raspberry Pi Radio Time Machine

The Raspberry Pi is a cheap, powerful single-board computer that's taking the maker world by storm. We'll show you how to build a Raspberry Pi-based audio player inside an antique wooden case, then upload some classic serials to revisit the glory days of radio. Who knows what evil lurks inside? You do!

SODA BOTTLE ROCKET

By Steve Lodefink

You don't have to be Burt Rutan to start your own rocket program. With a few empty soda bottles and some PVC pipe, you can build your own high-performance water rocket.

I've been a big fan of model rocketry since I built my first Estes Alpha back in third grade. Nothing is more exciting to a 9-year-old proto-geek than launching a homemade rocket.

But flying those one-shot solid-fuel rockets can burn a hole through a young hobbyist's wallet faster than they burn through the atmosphere, and with today's larger, high-powered rockets, locating and traveling to a safe and suitable launch site can require substantial planning and effort.

Instead, you can use 2-liter carbonated drink bottles to build an inexpensive, reusable water rocket. The thrill factor is surprisingly high, and you can fly them all day long for the cost of a little air and water. It's the perfect thing for those times when you just want to head down to the local soccer field and shoot off some rockets!

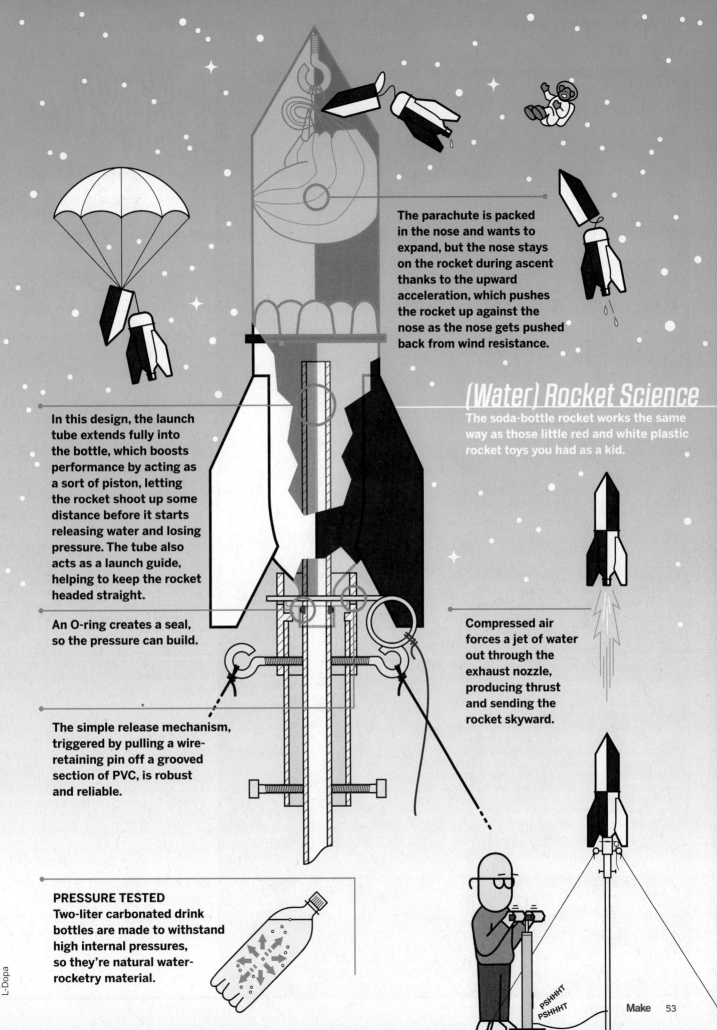

The parachute is packed in the nose and wants to expand, but the nose stays on the rocket during ascent thanks to the upward acceleration, which pushes the rocket up against the nose as the nose gets pushed back from wind resistance.

(Water) Rocket Science

The soda-bottle rocket works the same way as those little red and white plastic rocket toys you had as a kid.

In this design, the launch tube extends fully into the bottle, which boosts performance by acting as a sort of piston, letting the rocket shoot up some distance before it starts releasing water and losing pressure. The tube also acts as a launch guide, helping to keep the rocket headed straight.

An O-ring creates a seal, so the pressure can build.

Compressed air forces a jet of water out through the exhaust nozzle, producing thrust and sending the rocket skyward.

The simple release mechanism, triggered by pulling a wire-retaining pin off a grooved section of PVC, is robust and reliable.

PRESSURE TESTED
Two-liter carbonated drink bottles are made to withstand high internal pressures, so they're natural water-rocketry material.

PSHHHT
PSHHHT

Kirk von Rohr

BUILD YOUR SODA-BOTTLE ROCKET

1. BUILD THE LAUNCH TUBE

1a. Cut the tube. Use a hacksaw to cut the ½" PVC pipe to length. A 50" tube will make a launcher that's a convenient height for most adults to load from a standing position. The ½" Schedule 40 PVC pipe fits perfectly into the neck of a standard 2-liter soda bottle.

1b. Install the O-ring. Mark the O-ring position by fully inserting the launch tube into the type of bottle that you plan to use for your rockets. Locate the O-ring roughly in the middle of the bottle's neck. Use the edge of a file to cut a channel for the O-ring to occupy. Rotate the launch tube often while you work to maintain an even depth of cut, and be careful not to go too deep. Then slip the O-ring over the launch tube and seat it in the groove.

2. BUILD THE RE-LEASE MECHANISM

2a. Assemble the release body. Cut a 4" length of 1" PVC pipe and press-fit it into the 1" coupler. Cut squarely and deburr all PVC cuts with 120-grit sandpaper.

2b. Cut the release spring slots. Insert your bottle's neck into the release assembly and determine the distance of the bottle's neck flange from the end of the bottle. Mark the flange location on the 1" pipe coupler and use the hacksaw to cut a ³⁄₁₆" long slot on each side. These slots will hold the retainer/release spring.

2c. Attach bolts. Drill 3 evenly spaced holes through the release collar and release body together, and thread the 3 eyebolts into these holes. Similarly, drill 3 holes in the lower release body tube to accept the 3 hex bolts.

LAUNCHER MATERIALS

» **Schedule 40 PVC pipe, ½", 50" length** for the launch tube
» **PVC pipe, 1", 4" length** for the release body
» **PVC elbow fitting, ½"** for the end cap
» **PVC coupler fitting, 1"** for the release collar
» **PVC plug cap, ½"**
» **Rubber O-ring, 22mm outside diameter (OD), 16mm–17mm inside diameter (ID)**
» **Eyebolts, 2" (3)**
» **Hex bolts (3)**
» **Flexible vinyl tubing, ⁵⁄₁₆" OD × ³⁄₁₆" ID, 15' length**
» **Hose barb, ³⁄₁₆"**
» **Tire air valve**
» **Music wire, ⅛"** for the release spring
» **Nylon cord**
» **Small binder rings (3)** for stay clips
» **Small tent stakes (3)** for stays
» **PVC cement**
» **Bicycle pump with pressure gauge**

ROCKET MATERIALS

» **2-liter carbonated drink bottles (3)** We used Pepsi bottles.
» **Deli cup lid, 4"**
» **Fin material, such as balsa, thin plywood, or Plastruct sheeting**
» **Eyebolt, 2"**
» **Medium nylon washer**
» **Kite string**
» **Large garbage bag** for parachute material collar
» **Round hole reinforcement labels**
» **Quick-set epoxy**

TOOLS

» **Hacksaw**
» **Utility knife**
» **File, ⅛"**
» **Drill**
» **Locking pliers**
» **Sandpaper, 120-grit**
» **Thread-cutting taps and dies (optional)**

1a

1b

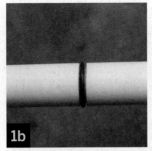

2a

2b

2d. Make the retainer/release spring.
Bend a piece of ⅛" music wire one and a half turns around a piece of scrap ½" pipe clamped into a vise. The spring should be roughly V-shaped.

Make a retainer clip for the spring by drilling 2 holes in a scrap of ½" pipe. The ends of the compressed spring will fit into these holes. This keeps the spring closed above the bottle's neck flange, holding the bottle in place.

Tie a 15' trigger line to the clip. At launch time, you pull the clip off with this trigger line, which allows the spring to open and the rocket to take off.

2e. Install the release assembly.
Slide the assembly into position on the launch tube, and tighten the 6 bolts evenly to keep it centered. To find the proper position, place a bottle on the launch tube and clip the release spring in place above the bottle's neck flange.

3. MAKE THE AIR HOSE

3a. Drill a ³⁄₁₆" hole in the center of the threaded ½" end cap, and press in the ³⁄₁₆" barb fitting.

3b. Thread the end cap into the elbow fitting and tighten it with a wrench. Using PVC cement, solvent-weld the elbow to the bottom end of the launch tube. The end cap is tapered, so it should require no Teflon tape or adhesive.

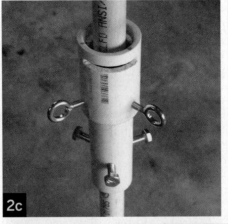

2c

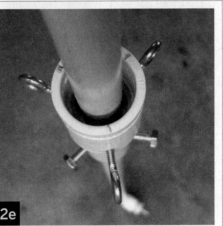

2e

2d

3a

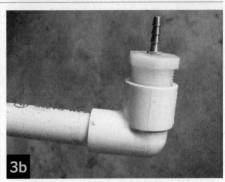

3b

3c. Use a utility knife to strip the rubber from the tire valve to one inch from the end. Insert the valve into one end of the ³⁄₁₆" flexible tubing.

3d. Push the other end of the air tube onto the barb fitting.

3c

3d

Steve Lodefink

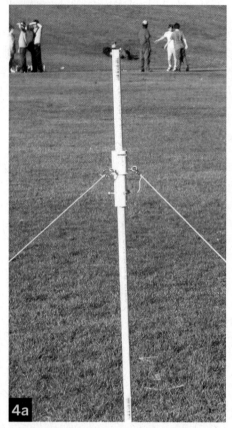

4a

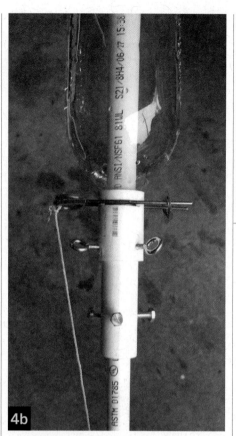

4b

4b

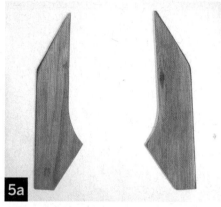

5a

4. SET UP AND TEST THE LAUNCHER

4a. Stake down the stays. The launcher is installed in a field using 3 stays, each consisting of a 72" length of light nylon cord. Stake one end of each line to the ground, and clip the other end of each stay to the eyebolts on the launcher.

4b. Pressure-test the launcher. Now is a good time to ensure that all the launcher's connections are airtight. Fill a bottle to the top with water (this way, if the bottle fails this pressure test, it will not explode). Quickly invert the bottle and slip it onto the launcher. A little Vaseline inside the neck will help the bottle make a seal against the O-ring. Squeeze the release spring into the slots in the release collar and clip it in place. Use the bicycle pump to pressurize the system to 70psi. If the pressure holds steady, all is well. Otherwise, fix any leaks and test again.

> TIP: If you can't fit the bottle over the O-ring, try these tricks: 1) bevel the inside of the bottle mouth by carving or sanding, 2) deepen the O-ring groove slightly, or 3) shave the O-ring with a razor to reduce its diameter.

5. ASSEMBLE THE ROCKET

Water rocket designs range from a simple finned bottle to those with elaborate six-stage systems with rocket-deployed parachute recovery and on-board video cameras. Ours is a painted single bottle affair with wood fins and parachute recovery. Chute deployment is by the passive "nose cone falls off at apogee" method.

5a. Cut 3 or 4 fins from a light, stiff material such as balsa, thin plywood, or Plastruct sheeting. Search the web for "water rocket fin template," and you'll find plenty of shapes to try.

Roughen the surface of the bottle with sandpaper where the fins will attach, and then glue the fins to the bottle with epoxy or a polyurethane adhesive such as PL Premium. Sand the leading edges smooth.

> TIP: Gluing on the fins at a slight angle will cause the rocket to spiral as it flies, adding stability to the flight. Try offsetting the base of the fin 4° from the top of the fin.

5b. Make the nose section by cutting off the neck and base of another bottle. Cut a 6" circle of material from a third bottle. Make a radial slit on the circle, fashion it into a nose cone, and cement it in place atop the nose section.

5c. Outfit the nose cone. When the cement is dry, turn the nose over and epoxy the 2" eyebolt to the inside tip of the nose cone. This bolt serves as a place to anchor the parachute shock cord. It also adds extra mass to the nose section, which will help to pull this section off as the rocket decelerates, exposing the parachute.

5d. Make the nose-stop. Cut the center from a 4" deli container lid, leaving only the outer rim. Cement the rim onto the rocket's lower "motor" section such that it allows the nose to sit loosely and straight on the rocket. This "nose-stop" will prevent the nose from being jammed on too tightly by the force of the launch, which ensures that the nose will separate off and deploy the parachute during descent.

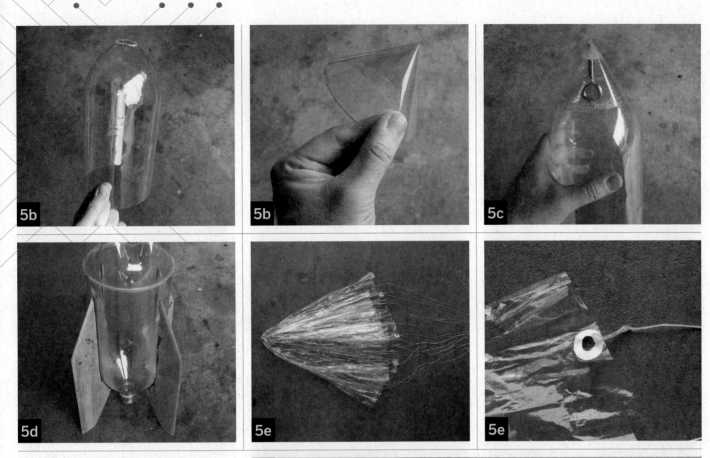

5e. Make a parachute canopy from a 36" or so circle cut from a large trash can liner. For best results, use 12 or more shrouds made from kite string. Apply paper reinforcement labels to both sides of the chute, where the shrouds attach, to keep the chute from tearing. Tie the loose ends of the shrouds to a nylon washer or ring to make the chute easy to manage.

> TIP: Ideally, a parachute's shrouds should be a bit longer than the diameter of the chute canopy.

5f. Epoxy a parachute-anchoring ring to the top of the rocket base and tie the parachute to the ring with a short cord. Cut a 4' connecting cord and tie it between the nose cone eyebolt and the parachute-anchoring ring. This cord will keep both halves of the rocket together during descent.

> TIP: Make sure the connecting cord is long enough to allow the parachute to completely pull out from the nose cone.

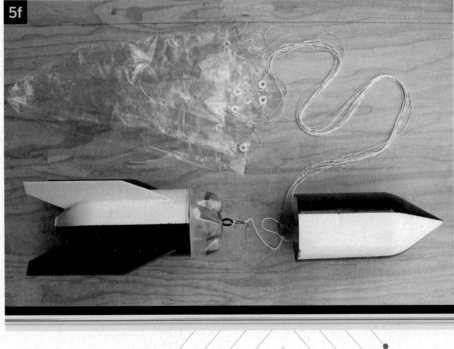

THREE, TWO, ONE, LIFTOFF!

Topher Lucas

SAFETY

Water rockets produce a considerable amount of thrust, and getting in the way of one could cause severe injury. Take the same common-sense precautions that you would when launching any type of rocket. Make sure that everyone in the area is clear of the rocket and aware that it is about to launch. Do a verbal countdown, or yell something alarming, such as "Fire in the hole!" just before you launch the rocket.

SELECTING A SITE

A well-built, single-stage water rocket is capable of flying several hundred feet into the air and drifting a considerable distance during descent. Less well-built rockets may choose to travel several hundred feet to the side. In any case,

you need to choose a launch site that is large and open enough to allow your rocket to wander a bit without getting lost in a tree, or on the roof of some Rottweiler's doghouse.

Big sports fields are the logical site choice for most of us, but if you are in a rural area, any wide-open space will work as a rocketry range. Be sure to take wind direction into consideration when deciding on which side of the field to set up the launcher.

OPERATION

1. Set up the launcher by clipping the 3 support stays to the launcher's eye-bolts. Take up any slack in the lines and stake the other ends to the ground, evenly spaced. Uncoil the air hose and tuck it under one of the tent stakes to

keep it from coiling. Attach the bicycle pump to the air hose.

2. Pack the parachute. Grab the center of the parachute canopy between your thumb and forefinger, and let it hang. Draw the chute through your closed hand to gather it, and then fold it into thirds, zigzag style. Lay down the parachute shrouds on the ground, and accordion-fold them back on themselves. Don't wrap the shrouds around the canopy; just slide the whole thing into the nose section and bring the 2 halves of the rocket together.

> **TIP:** Line the nose section with parchment paper or Teflon baking sheet liner to help the parachute deploy smoothly. A light dusting of talcum powder will also help keep the chute from sticking.

Watch a video clip of author Steve Lodefink and his 4-year-old son, Ivan, launching their soda-bottle rocket at makezine.com/go/sodabottlerocket.

3. Fill and set up the rocket. While holding the nose in place, turn the rocket over and fill it one-third full of water. Apply petroleum jelly to both the launch tube O-ring and the inside of the bottle's mouth to help it slip onto the O-ring. Hold the mouth of the rocket up to the launch tube. In one smooth motion, pivot the rocket up, slide it down onto the launch tube, and twist it back and forth, if necessary, to help it engage the O-ring seal.

> TIP: Tie the release spring to the launcher with some string to keep it from flying across the field and getting lost every time you launch.

4. Compress the release spring into the slots of the release collar, locking the bottle flange in place. Install the spring-retaining clip on the ends of the spring, and carefully run the trigger line back to your "ground control" area.

5. Launch! Jump over to the bicycle pump and bring the pressure up to about 70psi. When you're ready, clear the area, count down to zero, and pull the trigger line, releasing the spring and freeing the rocket.

If all goes as planned, your rocket will shoot upward, dispensing with its entire fuel load in less than half a second. Then it will begin to decelerate, and the nose will want to separate. As the rocket reaches apogee, the 2 halves will come apart, deploying the recovery chute and bringing the craft gently back to Earth, much to the excitement of the assembled crowd.

Experiment with different amounts of water and air pressure until you find the sweet spot that sends your rocket the highest. Don't exceed the amount of air pressure that your bottle is designed to withstand; 70psi seems to be about right for a standard 2-liter soda bottle.

ADVANCED DEVELOPMENT

Once you've tasted the joys of basic water rocketry, you'll inevitably want to improve and refine your rocket designs. If you want your rockets to fly higher, the best improvement you can make is to increase the volume of the rocket "motor." This is usually done by splicing or otherwise coupling 2 or more bottles into a single pressure chamber. There are also various schemes for building multi-stage rockets, as well as more elaborate parachute deployment setups.

There's an abundance of water rocket information available online. Here are a few sources to get you started:

Antigravity Research Corporation – ready-made water rocket components: antigravityresearch.com

Water rocket links: makezine.com/go/waterrockets

The Martinet Launcher, the basis for this project's launcher design: martinet.nl/articles/20050101

Steve Lodefink works as a software user interface and experience designer for the Walt Disney Company in Seattle.

SKILL BUILDER:
GETTING STARTED WITH SOLAR POWER

Ready to get off the grid, or just go green? Build your own 20-watt solar panels from inexpensive solar cells. Then learn the basics of designing a home solar power system and hook your panels up to storage batteries and inverters (and even the grid) in this 2-part series.

PVC Kids' Table and Stool

Learn to cut and fasten humble PVC pipe for building simple, sturdy structures. This small stool and table fit young kids perfectly — and they can scribble to their hearts' content on the dry-erase tabletop. There's even a stash for crayons and markers.

College Bike Trunk

Car drivers can lock things up while running errands, so why not cyclists? Build this sheet metal bike trunk, secure, weatherproof, and lockable. It'll hold two 1-gallon jugs of milk with room to spare — or if you're a college student, an ample haul of your beverage of choice.

COFFEE HACKS:
Bottomless Portafilter and PID Temperature Control

Espresso geeks spare no effort in their quest for perfection. Modify your espresso maker's filter holder for a tastier cup and more prized "crema" on top. Or add a precision temperature controller to the hackable Rancilio Silvia machine and get your shot truly dialed in.

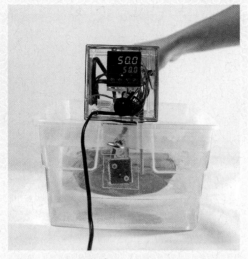

Sous Vide Immersion Cooker

In sous vide cooking, foods are vacuum-sealed and immersed in a temperature-controlled hot water bath to achieve optimal doneness and flavor. Commercial units cost thousands, but you can build your own smart, programmable immersion circulator for about $75 that heats water at temperatures accurate within 0.1°C. Get ready for incredibly rich eggs, perfect medium-rare steaks, and much more.

Fetch-O-Matic Dog Ball Launcher

Build a spring-loaded automatic launcher that whacks a tennis ball 25 feet, for about 50 feet of total roll. It runs on cordless drill batteries, so you can take it anywhere. You can even train your dog to drop the ball back into the hopper. Attaboy!

Homemade Fruit "Roll-Ups"

Fruit leather is a kids' snack trifecta — they love it, it's healthy, and — praises be — it doesn't melt or crumble all over clothes and car seats. It's also easy to make. Try this technique on peaches, cherries, or whatever needs rescuing from your fruit bowl this summer.

GARDUINO GARDEN CONTROLLER

Create an automated watering, light, and temperature control system to keep your plants happy. An Arduino microcontroller reads simple sensors and then waters the plants only when they're thirsty, turns on supplemental lights based on how much natural sunlight is received, and even alerts you if the temperature gets too cold.

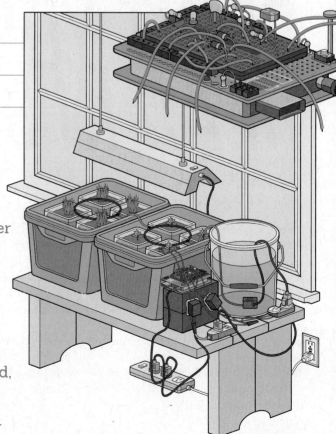

PRINT YOUR HEAD IN 3D!

By Keith Hammond

USE DIGITAL PHOTOS AND A 3D PRINTER TO MAKE A MINI PLASTIC REPLICA OF YOUR NOGGIN.

TOOLS

» **Computer with internet access and web browser**
» **MeshMixer software (optional)** free from meshmixer.com
» **3D printer and computer with printer's software (optional)** If you don't have access to a 3D printer, just send your 3D model to a service instead, and they'll print and mail it to you (see Step 10).

Here's a great project to get you started in 3D printing — create a 3D model of your own head and then print it out in solid plastic.

A 3D printer makes an object by squirting out a tiny filament of hot plastic, adding one layer at a time. That's why it's called "additive manufacturing." You send the printer a computer file that's a 3D model of something — an iPod case, a bike part, your head — then it prints out the object for you. These machines are becoming affordable for

schools, labs, libraries, and families, and there's lots of software out there for creating 3D files to print.

We chose Autodesk 123D software because it's free, it's web-based so you can use it from any computer, and amazingly, it lets you create a 3D model directly from digital photos. That way, you can do it all from home, and you don't have to get yourself scanned by a laser scanner.

When you're done making your 3D model, you can take it to a makerspace

where they have a 3D printer, or you can send it out to a service and they'll print it and mail it right to your home. We printed our heads on an Ultimaker printer, using Cura as the printer software. It was easy!

Imagine what else you could 3D-print with these tools. Instead of printing your head, why not replicas of buildings or sculptures at an art museum? Or you could make models of your pets, your car — almost anything you can capture in photos.

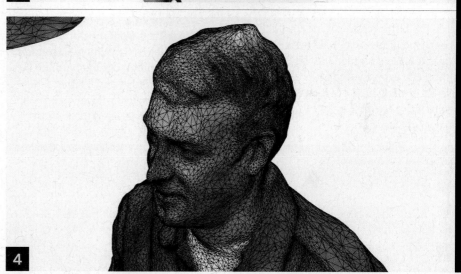

1. REGISTER WITH AUTODESK 123D

Go to 123dapp.com and create a free account. For this project, we'll use the web app for 123D Catch. It stitches your digital photos together into a 3D model.

Autodesk recently updated 123D Catch so you can 3D-print your head two ways: either send your model out to be printed for you, or download it so you can print it yourself. (There's a powerful desktop PC version of 123D Catch, but you won't need it for this project.)

2. TAKE DIGITAL PHOTOS OF YOUR HEAD

You'll want a friend's help with this part. You can use a cellphone camera or a nice DSLR — the better the camera, the better 123D Catch will work. Shooting in full shade works best.

Sit still while your friend snaps 30 to 40 photos of your head, in 2 separate loops moving completely around you — one lower loop, and one higher loop where the top of your head is seen clearly. This will prevent unwanted holes in your head where the software is missing part of the scene. For best results, make sure your head fills most of the frame.

If you're going to stick out your tongue or make a face, ask your friend to work fast so you can hold your expression. But remember to keep the camera still and focused when snapping each photo, because blurry images may confuse the software and cause weird horns on your head.

3. CREATE A NEW CAPTURE

In 123D Catch, upload all of your head photos. In the Model Resolution pull-down menu, select High (For Fabrication). Give your model a name and click Create Model.

Autodesk's computers will automatically stitch all your photos together to make a 3D model, and then put the model in your My Projects section.

4. OPEN YOUR 3D MODEL

You're looking at yourself as a 3D model! It's got a realistic texture, like your original photos. You can Dolly, Pan, and Orbit to move your view around, by using those 3 buttons on the right-hand toolbar.

On the same toolbar, select Material & Outlines to see the 3D mesh that's underneath the texture. Cool!

5. EDIT YOUR 3D MODEL

My 3D model had a crazy horn on the back of my head, maybe because we took some photos that were blurry or too far away. It also captured background elements that we don't want to print. To remove major unwanted features, use the Select Faces tool to highlight them, and then Delete them. (Or highlight your model, click Invert Selection, and delete everything but your model.) To snip a horn from your head, use the Delete & Fill tool, then use the Smooth Brush to round it off. Trim your model to size and save it under a new name.

6. MAKE IT "WATERTIGHT"

Click on Inspect Model and Cap All to automatically repair any holes.

The bottom of your model will be closed now, but it might be an extended blob. For best results on the 3D printer, your model should be flat on the bottom. Click on Plane Cut Model, then drag and/or rotate the plane to where you want to slice the bottom off your model. Click on Apply and your model will have a flat bottom. Re-save your model to My Projects.

You can export your model as an STL file for printing now, or fool with it some more using MeshMixer software as shown in Step 7.

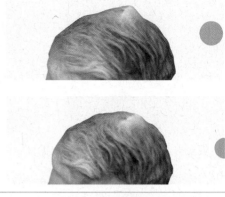

7. EMBELLISH IT (OPTIONAL)

MeshMixer (free from meshmixer.com) is a powerful tool for editing 3D models and merging them together. Autodesk recently acquired it, and it's frequently updated. Before using it, we recommend that you watch the video tutorials at youtube.com/user/meshmixer.

For a quick-and-dirty pedestal, open your STL file in MeshMixer. Select the whole mesh (Ctrl-A or Cmd-A), then select Edits → Plane Cut to slice off the bottom. Select the bottom face of the model and click Edits → Extrude. In the Tool Properties bar on the right, set the EndType to Flat. Then click and drag the Offset bar to extend your model, creating a simple pedestal that's perfectly flat. Click Accept and save a new STL file.

To merge your head with a fancy pedestal, start with a 123D mesh that's still open on the bottom. Select the whole mesh, choose Edit → Convert to Part, and click Accept. Look at the Parts bar on the left: your head is now a "part" you can merge with other parts. Now import an STL file of a pedestal — I like the pawn from Mark Durbin's Column Chess Set (thingiverse.com/thing:19659). Open it in MeshMixer, scale it to match your head, then drag your head onto it to merge the two. If it doesn't work the first time, try Edit → Remesh. Save as a new STL file.

You can do lots more with MeshMixer. Put bunny ears on your head, or stick octopus tentacles on it, or make yourself a two-headed monster. Or put your head on a Pez candy dispenser!

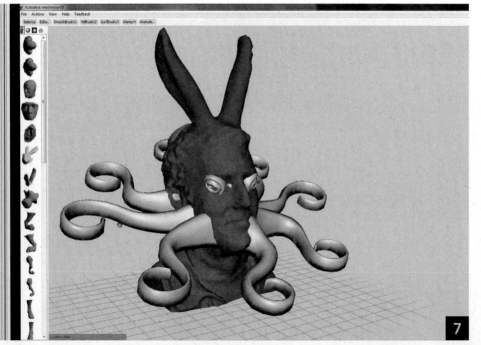

8. SHARE YOUR MODEL (OPTIONAL)

When your model is done, click Publish to Gallery. Now anyone can open it in a web browser and play with it. (If you're using the desktop version of 123D Catch, you can make a video animation and send it straight to YouTube.)

9. SAVE YOUR FINAL MODEL AS A PRINTABLE FILE (OPTIONAL)

To print your own head, you need a copy of your 3D model in a format that 3D printers can understand. Export your model from 123D Catch (or from MeshMixer) as an STL file. (Fun fact: STL stands for *stereolithography*, which is a different type of 3D printing.)

If you're sending your head out to be printed by Autodesk, you can skip this step.

10. 3D-PRINT YOUR HEAD!

Find a place where you can use a 3D printer (see directories at makerspace. com and hackerspaces.org). Bring your STL file on a thumb drive.

We printed the heads on page 62 on the Ultimaker in the MAKE Labs, which we like because it's fast and accurate — and because you can buy it as a kit and build it yourself. We've also had good success with a MakerBot Thing-O-Matic.

First, you'll open your STL file in the 3D printer's software, which tells the printer exactly where to make trails with the hot plastic to build up your object. For example, if your MakerBot uses ReplicatorG software, import your STL file, center the model and put it on the platform, then scale it to your desired size. Next, choose Generate GCode, select the default print profile, and check the Use Print-O-Matic checkbox. Now hit Print.

Watch in amazement as your head materializes before your eyes.

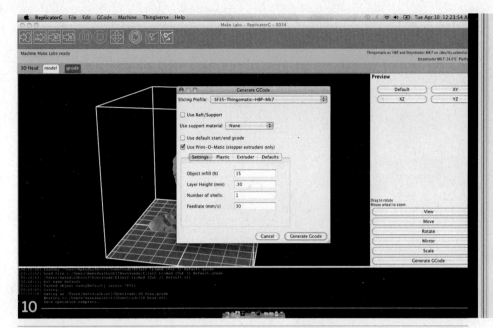

If there's no 3D printer close by, that's OK — lots of service companies will print out your 3D model for you. At 123dapp. com, select your project and click Fabricate → 3D Print to send your file to Autodesk's digital fabrication service and receive your 3D-printed plastic head in your mailbox. It only costs about $10 for a 3"-tall head.

Or try sending your file to Shapeways (shapeways.com) or Ponoko (ponoko.com), or in Europe, try Sculpteo or i.materalise. Some of these services will even print out your head in ceramic, glass, steel, silver, gold, or titanium!

123D tutorial videos from Autodesk: youtube.com/123d **and** 123dapp.com/catch/learn

5-Minute Foam Factory

Build this easy hot-wire foam cutter and start carving styrofoam into any shape you imagine. You'll learn to double- and triple-cut 3D shapes, spin a compound-curve cone, and cut a stack of foam sheets to make a blizzard of snowflakes — plus make simple tools for circle cutting, bevels, and more.

$30 Micro Forge

Always wanted to try blacksmithing? Grab your hammer and tongs and build this mini forge from a propane torch and a firebrick. It'll heat iron to white-hot so you can forge your own nails, chain links, and other small iron parts for carpentry, decorative hardware, fantasy, or historic reenactment.

SKILL BUILDER:
CNC PANEL JOINERY

Suddenly the world is full of people designing models, project enclosures, sculpture, furniture, and all kinds of other cool stuff to be slotted together from parts made on laser cutters and CNC routers. Here's a clever bag of tricks for making interlocking, self-aligning, and demountable joints in flat stock.

Kitchen-Floor Vacuum Former

From coffee lids to stormtrooper costumes, vacuum-formed plastic is everywhere. Build this simple vacuum box, hook it up to your vacuum cleaner, and you'll be molding parts like the pros, using simple suction to conform warm plastic tightly to whatever mold you choose.

SKILL BUILDER:
UNDERSTANDING BASIC WOODWORKING TOOLS

For a lot of makers, nothing satisfies like sawdust and shavings. Get your hands on the five basic hand tools — the hammer and chisel, hand plane, handsaw, and clamp — and learn how to sharpen your edges to make the shavings fly. Then move on to squares, snap lines, and other layout tools, then step up to power tools.

Car Battery Welding

Welding is a glorious, mystery-infused, thoroughly badass way to stick things together. Welders are dirtier, tougher, and sexier than other makers, and the things they build are big and strong and hold our world together. Build your own welding rig with three car batteries and a box of cheap welding rods.

Hovercraft Shop Vac

Casters never roll where you tell them to. Modify your drum vacuum to be self-levitating so it will float obediently behind you on a cushion of air.

Japanese-Style Workhorses

Use a simple mortise-and-tenon joint to make these fine-looking shop horses that'll last a lifetime. They're a good beginning joinery project without any metal fasteners — instead, they use the drawbore style of mortise and tenon, which is secured by a wooden pin that draws it tight and makes it look great.

Sam Murphy

TABLETOP BIOSPHERE

By Martin John Brown

The Tabletop Shrimp Support Module (TSSM) is a fun demonstration of the ecological cycles that keep us alive — and an enticement to muse on everything from godhood to space colonization.

When my 7th grade vocational aptitude test came back stamped "Forester" instead of "Astronaut," I knew the test-makers had screwed up. Sure, I liked sitting in streams, and peering down those creepy holes by the roots of old trees. But I also knew that someday the whole frickin' park would be flying through space. Hadn't anyone else seen *Battlestar Galactica*?

Now we know that space colonists are just as likely to be muddy ecologists as hotshot flyboys — the kind of people who assemble ecosystems instead of engines. Today's pack-it-in, pack-it-out life support is impractical for long, manned missions, but in the future, regenerative systems could provide years' worth of food, air, and water while processing human waste. It's recycling and reuse on a radical scale, light years beyond anything pitched by those hairy guys down at the co-op.

Here's a mini version of this dream, a sealed system that supplies a freshwater shrimp "econaut" with food, oxygen, and waste processing for a desktop journey of 3 months or more.

TABLETOP SHRIMP SUPPORT MODULE: HOW IT WORKS

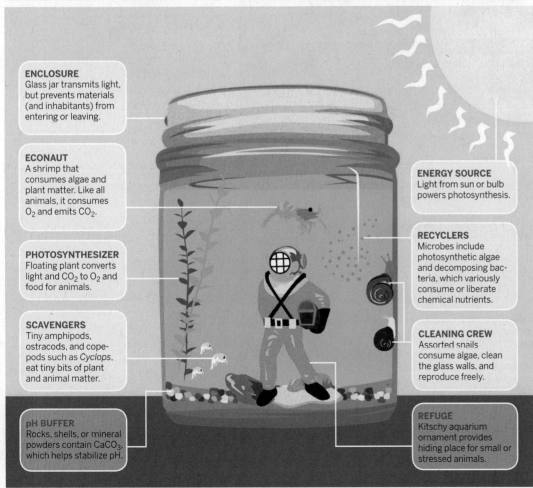

ENCLOSURE
Glass jar transmits light, but prevents materials (and inhabitants) from entering or leaving.

ECONAUT
A shrimp that consumes algae and plant matter. Like all animals, it consumes O_2 and emits CO_2.

PHOTOSYNTHESIZER
Floating plant converts light and CO_2 to O_2 and food for animals.

SCAVENGERS
Tiny amphipods, ostracods, and copepods such as *Cyclops*, eat tiny bits of plant and animal matter.

pH BUFFER
Rocks, shells, or mineral powders contain $CaCO_3$, which helps stabilize pH.

ENERGY SOURCE
Light from sun or bulb powers photosynthesis.

RECYCLERS
Microbes include photosynthetic algae and decomposing bacteria, which variously consume or liberate chemical nutrients.

CLEANING CREW
Assorted snails consume algae, clean the glass walls, and reproduce freely.

REFUGE
Kitschy aquarium ornament provides hiding place for small or stressed animals.

Our Bottled-Up World
On Spaceship Earth, little goes in or out except light and heat, and all organisms live off each other's waste, whether it's oxygen from plants or feces from animals. Our world is bottled up.

Ecologists have often scaled down these processes, creating sealed aquariums for research. Meanwhile, space scientists have searched for organism and machine combinations that could cooperate to support humans in a space colony.

The TSSM's basic principles come from ecologist H.T. Odum, but many details derive from the Autonomous Biological System (ABS), a sealed aquarium invented by Jane Poynter, which has returned healthy from extended trips on the space shuttle and the Mir and ISS space stations.

The Cast of Waterworld
In our TSSM, the "econaut" we imagine ourselves in the place of is a shrimp. We encourage photosynthesis and waste processing with abundant light and vascular plants, and we limit oxygen demand by constraining animal biomass and algae-fertilizing nitrate and phosphate. Protection against chemical spikes comes from pH buffers.

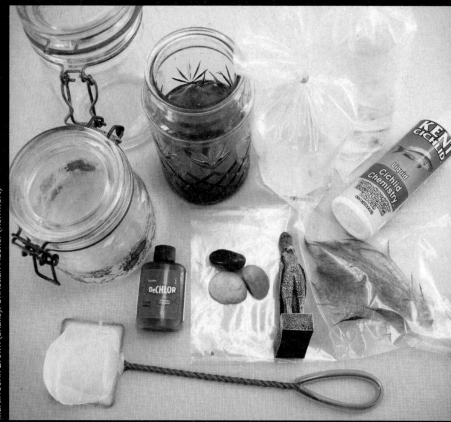

Martin John Brown (snails); Christian Fischer (hornwort)

MATERIALS

» **1-quart glass canning jar** Don't use plastic; it may bleed air.

» **Clear bottles or plastic containers** for sampling and a "holding tank"

» **Tap water**

» **Small river rocks** just enough to cover the jar bottom. Rocks piled too thick let muck and algae build up where snails and shrimp cannot eat them.

FROM AN AQUARIUM STORE

» **Tap-water dechlorinator**

» **Aquarium ornament(s) or other glass or ceramic obstacle(s)** Seashells also are nice, and supply extra calcium carbonate.

» **Fine fishnet or kitchen strainer**

» **Freshwater minerals** such as "Kent Freshwater" or "cichlid salts" These are essential trace nutrients.

» **Amano shrimp (1)** (*Caridina multidentata*) an algae-eater with a reputation for tolerating high pH

» **Snails (4)** of assorted species smaller than 1cm each.

» **8 stem inches of hornwort** (*Ceratophyllum demersum*)

» **2"×2" piece of duckweed (Lemna).** You can also collect this from a local pond.

[NOT SHOWN]

» **1Tbsp powdered calcium carbonate** This is your primary pH buffer.

FROM A LOCAL POND

» **Assorted amphipods (2–8)** These are tiny crustaceans; try to collect 8, but you can use fewer.

» **1 or 2Tbsp pond sludge** hopefully containing copepods and ostracods (even tinier crustaceans), bacteria, microalgae, etc.

NOTE: **Aquarium fish, shrimps, and snails may be invasive and destructive if released into the environment, so boil or freeze them after the experiment. Or keep them living in an aquarium environment.**

CREATE YOUR BIOSPHERE

1. GATHER THE AQUARIUM SUPPLIES

1a. Visit an aquarium store for the materials listed on this page. While you're there, ask them how to dechlorinate local tap water for aquarium use.

NOTE: **The store staff might not believe that your Tabletop Shrimp Support Module will work. Make nice anyway.**

1b. At home, dump your shrimp, snails, hornwort, duckweed, and the water they came in into an open "holding tank." I use a plastic Tupperware or yogurt container. Add some dechlorinated tap water to keep everything comfortable (alive).

2. COLLECT THE POND LIFE

Go to a local pond. Spring and summer are best. Bring a net or bottle (or other container), and visit during late afternoon. That's when the pH is higher, like that of your TSSM.

2a. Find a good, shallow area of the pond to collect your goodies. If you see duckweed, water lilies, or other vascular plants, try near there. I've done well in areas with a mixture of substrates, like sand, rock, and decaying wood.

2b. Drag your bottle or net through mud, rocks, and half-submerged plants. Examine your take for shrimp-like creatures 1mm–10mm long. These are probably amphipods; collect up to 8 of these if you can. You need to look aggressively, getting into the muck and shaking bits of plant away. Then collect 1 or 2Tbsp of pond sludge from the pond bottom, which should contain some nearly microscopic copepods and ostracods. Back home, dump your pond samples and sludge into the holding tank.

3. BOTTLE IT UP

3a. In a new container, whip up a gallon of NPFW (nitrate-poor fresh water). This is tap water, dechlorinated and supplemented with your freshwater mineral mix (follow package directions).

> **NOTE:** Waters from the aquarium store and pond are probably loaded with algae and algae-supporting nitrates, which will lead to algae takeover. Diluting with NPFW helps prevent this.

3b. Thoroughly rinse your "fixtures" — quart canning jar, ornaments, rocks, etc. — with NPFW.

3c. Fill your jar halfway with NPFW, and transfer all the ingredients to the jar, except for calcium carbonate powder, if used: shrimp, snails, hornwort, duckweed, amphipods, sludge, ornaments, rocks, seashells. Use the quantities listed. Do not put in extra animals or sludge, or otherwise mimic a traditional aquarium. What makes this system work is its sparseness.

3d. Fill the remaining volume of the canning jar with NPFW, leaving 1" or 2" of airspace at the top. If you have calcium carbonate, add it last, and note that it will cloud the water for hours.

3e. Say a little prayer as you tighten the cap on the jar.

3f. Your biosphere is complete! Place it in a spot with a fairly consistent temperature (70–80°F) and 12–16 daily hours of moderate light. Standard room lighting is too dim, and direct sun is too much. A bright north window or a 50W bulb a few feet away are both good, but watch the temperature.

Martin John Brown (martinjohnbrown.net) is a writer and researcher specializing in environmental and historical topics. He really likes the blog bottleworld.net.

1a

2b

Martin John Brown (aquarium store & pond)

3a

3a

3a

3c

3d

3f

ENJOY YOUR BIOSPHERE

Maintaining the TSSM is a joy. There's no feeding or fiddling with parameters. Just observe and philosophize. Get enchanted with your econaut shrimp, casting its antennae in slow looping rhythms. Watch the snails cruise the glass like silent Sumo wrestlers on night patrol. Zoom in on the tiny creatures oozing out of the muck. They are the bottom of the food chain, the disassemblers of the dead.

FROM LEFT TO RIGHT: Amano shrimp chills upside down; snail grazes on algae; snail-on-hornwort action.

There's never been another world like this one. In a way, you're God! Which might bring on some curious emotions if something goes awry. Multispecies assemblages like the TSSM are never 100% reliable. Your econaut might die mysteriously. Or you might observe signs of stress: shrimp that molt and then shrink instead of grow, or carnivory among normally vegetarian shrimp or snails. Hard questions arise. Was it right to start this world? Will you intervene, or abandon your creations to a sealed fate?

Life inside such tight ecological loops is rarely a cakewalk, and this begs some questions. Does closed-system sustainability simply emerge as you scale things up? Or is there something about the Earth and its milieu of flux on flux that we've failed to understand so far? Might our increasingly crowded planet, with a rising rate of extinctions, start resembling a laboratory microcosm? And for those with sci-fi dreams, could living on Mars be little more than desperate farming?

But if ecosystems engineering makes progress, we have hope. Mark Kliss, chief of the Bioengineering Branch at NASA's Ames Research Center, envisions extraterrestrial life support systems that provide a high quality of life, with a big contribution from automation. Machines and software could monitor conditions and energy inputs, nudging ecological feedback loops away from mutual parasitism and into productive symbiosis.

It's a vision our environmental movement might consider. The thing that finally allows people to live in balance with nature might be technology, the force that once seemed most opposed to it.

BC Anna (snail)

BEYOND SPACESHIP EARTH

At least 5kg of food, water, and oxygen must be lifted into space for every person-day spent on the International Space Station, relates NASA's Mark Kliss. For human habitation on the Moon, Mars, or elsewhere — stays of hundreds or thousands of days — that adds up to an unworkable ball and chain.

That's why Kliss and others are trying to replicate the closed-system sustainability of Spaceship Earth. Academics have long built "closed ecosystem" models for streams or lakes to investigate subjects like carbon cycling and population dynamics.

For potential space travel, the conditions are far more constrained. Species may be mixed in ways never seen in nature, but must include the target species, humans.

American and Russian space scientists have been working on the problem since the 1960s. Early Russian tests were brutally simple: one guy climbed into a cask with little more than a light and a bucket of photosynthetic algae, to stumble out 24 hours later, alive and stinking. Progress has been slow, and no bioregenerative systems have yet been used in space for human life support.

Research has followed two paths. Space agencies have focused on highly engineered systems that include just a few well-understood species and fully account for their chemical products and needs.

Projects like Biosphere 2 (and the TSSM project here), however, take a more top-down approach. Thousands of species were imported to Biosphere 2's fantastic glass structure in the Arizona desert, and assembled into new forests, farms, and "oceans." By the time eight jump-suited "econauts" were sealed in, in 1991, it was a publicity juggernaut.

Over the next two years — the duration of a Mars expedition — the econauts met the recycling challenge, surviving very largely on regenerated air, food, and water. But their elaborate menagerie suffered a hard shakeout. Oxygen declined to dangerously low levels, and food became scarce. Extinctions were rampant and,

"Econaut" Jane Poynter working within Biosphere 2.

critically, included all the pollinating species.

Life in Biosphere 2, that questing ecological utopia, wasn't sustainable. When ecosystems are sealed off, it's *Escape from New York*. Systems must balance locally, and an ecological shakeout ensues. The community that emerges may be strange and new, or as dismal as pond scum. Even with our TSSM, you can follow the same recipe to bottle up more than one tabletop biosphere, and things will evolve in different directions.

As Kliss philosophizes, closed ecosystems tread a fine line between symbiosis and mutual parasitism. Will the inhabitants help each other survive, or eat each other alive?

Resources at makezine.com/10/ biosphere

JELLYFISH TANK

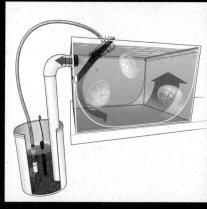

A tank of healthy jellyfish is hypnotically beautiful. Unfortunately, jellies can't live in a regular aquarium because they get sucked into the filtration pumps or trapped in "dead spots" at the edges. This special design creates a laminar flow that keeps jellies safely suspended in the middle of the tank.

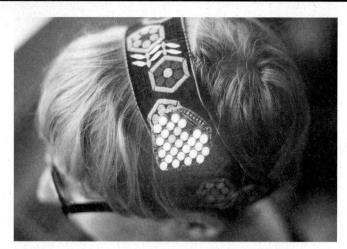

Beating Heart Headband

Build a pulse-sensing headband that flashes an LED array in time to the beating of your heart. You'll scratch-build your own perfboard Arduino, assemble the Open Heart LED display, and learn how to use the AMPED pulse sensor. It's the next best thing to wearing your heart on your sleeve.

Home Molecular Genetics

Working directly with DNA isn't only for the labs of CSI, agribusiness, and headline-grabbing research institutions. It's basic chemistry, but it uses the molecules of life. You can even do it at home. This project explains how you can isolate and even "fingerprint" some of your own DNA.

Lensless Microscope

Inside your webcam is an imager chip with thousands of sensors, each much smaller than a hair. Replace the lens with an LED, place a sample directly on the chip, and you have a lensless microscope capable of streaming live video of plankton or other microorganisms to any display.

Geiger Counter

This DIY Geiger counter clicks and flashes an LED each time it detects a radioactive particle. It works with most common Geiger–Müller tubes, and easily connects to radiation rate meters, data loggers, true random number generators, and the Radiation Network's GeigerGraph software. Share your radiation readings with the whole world!

BACKYARD BIODIESEL

Biodiesel is vegetable oil (often used) that's been chemically converted into a fuel thin enough to spray through a diesel fuel-injection system. No special equipment is needed to "cook" a small batch of biodiesel that will work in any diesel engine, from a model airplane to the family car.

BLEACH SHIRT STENCIL

NOT YOUR FATHER'S TIE-DYE

By Bre Pettis

You want a design on a T-shirt and you want it *now*! Sometimes the best designs come out of desperation. Here's how to put a cool design on a T-shirt ASAP!

YOU WILL NEED
» Dark T-shirt
» Gloves
» Safety goggles
» Bleach
» Spray bottle
» Piece of cardboard
» Pen or pencil
» Marker
» Paper
» Scissors
» Sink
» Water
» Spray-adhesive tack (optional)

1. PREPARE

Put on gloves and safety goggles. Mix up a batch of 50/50 bleach to water solution in a spray bottle. Label it, and draw a skull and crossbones on the bottle so nobody mistakenly thinks it contains water. Then put a piece of cardboard inside the shirt so the bleach doesn't bleed through.

2. DESIGN

Draw a stencil design on paper and cut it out. Just remember that it's going to be a stencil, so it should be designed so it can be filled in with color (or bleach). Bleach can bleed into the shirt, so keep it simple and be bold. When you're done, place the design on the shirt. I put some spray-adhesive tack on the back of the stencil so it would stick to the shirt a bit.

3. BLEACH

Set your sprayer to a fine mist and spray the bleach solution over the stencil. Watch the color disappear before your eyes.

4. RINSE

If you leave the bleach on too long, it'll make the fabric disintegrate, so when it's getting close to the right color, dunk the shirt in a sink full of water and rinse until it doesn't smell like bleach. Dry it out, and you're good to go — your T-shirt design is done!

Bre Pettis makes things that make things. He is the CEO of MakerBot Industries.

CUSTOM CONVERSE UPPERS

Originally created in 1917 as basketball shoes, "Chucks" (nicknamed after basketball player Chuck Taylor) quickly caught on to become one of the most popular shoe brands of all time. It's their playful shape, easy-to-cut canvas, accessibility, and relative low cost that make these sneakers fun to hack. So try this fabric swap or knitted Chuck project and consider yourself a Converse crafter.

War and Peace Bookend

Turn war into peace by converting a children's action figure into a peace-loving bookend. Or if you don't have a soldier, try using an old space ranger (you know the one we're thinking of) or any other large toy figure.

Shiny Globes of Mud

Everyone enjoyed playing in the mud as a child, right? Well, that's what you get to do with *hikaru dorodango* ("shining mud ball" in Japanese). Except this time, you don't just wash the mud off and forget about it. Instead, you create a beautiful and unique shiny ball of mud.

JAM JAR LANTERNS

Lighting makes an event, and candles transform the mundane into the extraordinary. These glass jar lanterns are beautiful hanging from a tree, or from chain across a deck. Pick any jar you like, then bend a snug-fitting cage of wire on a custom-made jig. The technique makes it easy to mass-produce a large number of lanterns in a short time.

Needle Felting

Needle felting is the art of sculpting wool with a barbed needle. As a medium, carded-wool batting can be manipulated into any shape. Infinite varieties and colors of wool are available to make flowers, dinosaurs, cats, dogs, robots, jewelry, or any sculpture. The supplies are inexpensive, and the techniques are simple and fun.

Build an Inkle Loom

Weaving is a great craft, because it's meditative and challenging, practical and artistic. Build this easy, inexpensive inkle loom to weave a belt or a unique strap for your messenger bag or camera. The slot-and-peg tensioning rod allows you to move the continuous warp through the loom and weave the entire length, giving you eight feet of warp that you can weave all in one go without stopping.

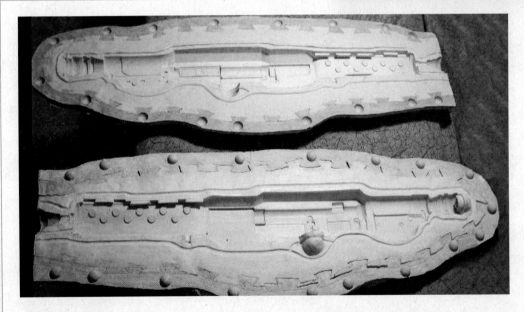

SKILL BUILDER:
MOLDMAKING

Molds can be made of rubber, steel, clay, brass, wood, plastic, lead, and even sand. Adam Savage of MythBusters fame shows you how to replicate objects of any size by casting them in rubber, which is a reliable technique that's employed throughout the film and special effects industries, as well as the jewelry industry.

STROBOSCOPE

PLAY WITH SPACE AND TIME
By Nicole Catrett and Walter Kitundu

FREEZE FRAME

A spinning disk with a single slit lets your camera see serial glimpses of a moving subject — and record them all in a single image.

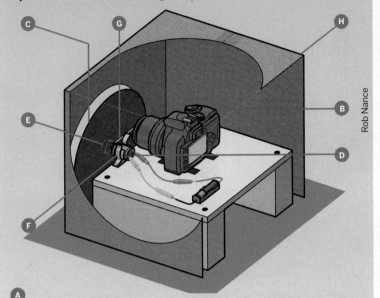

Rob Nance

Steve Hoefer

A The subject moves through space. Lights and a black background ensure that each successive image is sharp and distinct.

B A camera is aimed at the subject.

C A black strobe disk covers the camera's lens, preventing light from entering.

D A battery-powered DC motor spins the strobe disk.

E A cork in the center of the disk makes it easy to mount and remove from the motor.

F The motor is mounted to a base with a broom-holder clip.

G A slit in the spinning strobe disk lets a bit of light into the lens as it passes over. When the camera is set to a long exposure time, each slit passage allows a new image to overlay onto the camera's image sensor.

H A cardboard camera hood reduces ambient light that can cloud the captured images.

MATERIALS

» **Cork** from a wine or champagne bottle
» **Battery holder, 1xAA and battery** RadioShack (radioshack.com) part #270-401, $1
» **Motor, small, 1.5V–3V DC** RadioShack #273-223, $4
» **Alligator test leads (2)** RadioShack #278-001, $10 for a set of 4
» **Foam sheet ("Foamies"), 2mm×9"×12"** Craft Supplies For Less #F2BB10 (craftsuppliesforless.com), $4 for a pack of 10
» **Broom-holder spring clip with mounting screw** McMaster-Carr #1722A43 (mcmaster.com), $11 for a pack of 10. You can also use a zip tie.
» **Zip tie**
» **Digital SLR camera with manual focus that can take long exposures** Some good options are the Nikon D40, D3000, and D5000, and the Canon EOS Rebel Ti, XTi, XS, or XT. You might be able to use a point-and-shoot camera, but it needs to have a manual focus and exposure mode.
» **Plywood, ½" thick, 10"×10"**

» **Lumber, 2×4 , 10" lengths (2)**
» **Wood screws, 1¼" (4)**
» **Black fabric, enough to work as a backdrop** A queen sheet is a good size, and black cotton works well. Avoid anything shiny or reflective. You could also use matte-finish black paint for your backdrop.
» **Clamp lights (2 or more)** Home Depot #CE-303PDQ (homedepot.com), $14 each
» **Velcro tape, 3½" strips** OfficeMax #09015086 (officemax.com), $5 for a pack of 10
» **Construction paper or cardstock, black (1 sheet)**
» **Cardboard box** approximately 12" square

TOOLS

» **Drawing compass and pencil**
» **Ruler or calipers**
» **X-Acto knife**
» **Scissors**
» **Hot glue gun**
» **Sewing needle, large**
» **Screwdriver, Phillips head**
» **Drill and drill bits**

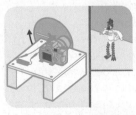

Make a mechanical strobe with a toy motor and construction paper, pair it with a digital SLR camera, and take stunning photographs of objects in motion.

We were inspired to play with stroboscopic photography after seeing photographs taken by 19th-century French scientist Étienne-Jules Marey. In the 1880s, Marey invented a camera with a rotating shutter that could capture multiple images on a single photographic plate. He used this camera to study locomotion in humans, animals, birds, sea creatures, and insects.

Marey used clockwork mechanisms and photographic plates for his contraption, but you can make a much simpler version with a slotted paper disk, a toy motor, and a digital camera. The camera is set to take long exposures while the slotted disk spins in front of its lens. Each time the slot spins past the lens, the camera gets a glimpse of your subject and adds another layer to the image. The resulting photograph is a record of your subject moving through space and time, and these images often reveal beautiful patterns that would otherwise be invisible to us.

BUILD YOUR STROBOSCOPE

1a

1b

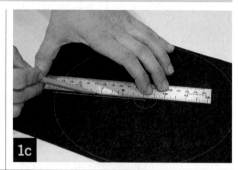

1c

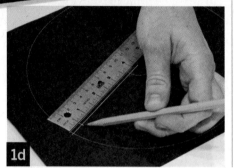

1d

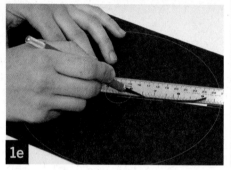

1e

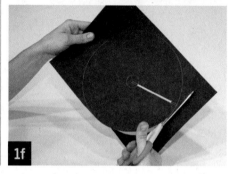

1f

1. MAKE THE STROBE DISK

1a. Use a compass to draw an 8½"-diameter circle on black construction paper or cardstock; apply enough pressure with the compass point to mark the circle's center.

1b. Measure the diameter of your cork. Use the compass to draw a circle of that diameter at the center of the larger circle. This will come in handy later when you mount the strobe disk on the motor shaft.

1c. For the strobe disk slot, use a ruler to draw a line from the center mark to the outer edge of the big circle. Then draw a second line ⅛" away from and parallel to the first line.

1d. Make a perpendicular mark across the 2 slot lines ½" in from the edge of the big circle. The slot will run from this mark to the closest edge of the inner circle.

1e. Cut out the slot with an X-Acto knife, using the ruler as a guide to make the cuts clean and straight.

1f. Use scissors to cut out the large circle.

2. ATTACH THE DISK TO THE MOTOR

2a. Use an X-Acto knife to slice a round section from the cork, ½" thick.

2b. Hot-glue the flat side of the cork section to the center of the strobe disk, using the small center circle you drew as a guide.

2c. Use a large sewing needle to pierce the center of the cork through the disk. This will create the pathway for the motor shaft.

2d. Push the strobe disk and cork onto the motor shaft, with the cork facing out. This should be a snug fit.

2e. To test the motor and strobe disk, use alligator test leads to connect the toy motor to your AA battery holder. The strobe disk should begin to spin.

You now have the fundamental parts of your stroboscope. Peer through the contraption at anything moving, and the scene will turn into an old-time movie. Watching your friends dance will be a whole new syncopated experience. Also try looking at a vibrating guitar string or a stream of water.

3. BUILD THE BASE

3a. Cut a 10" square piece of plywood for the base platform. Position a broom-holder clip (screw-hole down) with its long side parallel to an edge of the plywood, spaced approximately ¼" in from the edge and 2" in from a corner. Screw the clip down tight. (If you're using a zip tie instead of a broom clip, see Step 3c.)

3b. Remove the strobe disk from the motor and set it aside. Wrap a thin foam sheet (cut to fit) around the motor and pop it into the broom holder, shaft pointing out. The foam keeps the motor from vibrating.

3c. (Alternative) If you don't have a broom holder, you can use a zip tie. Drill 2 parallel holes (each large enough to

2a

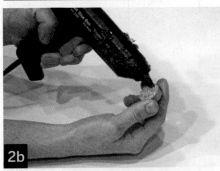

2b

2c

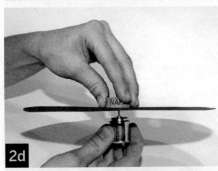

2d

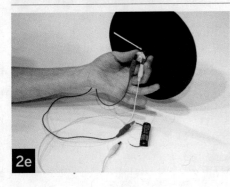

2e

accommodate the diameter of your zip tie) ¼" in from the edge of the plywood and 2" and 3" from the corner, respectively, and pass the zip tie through the holes and around the motor to secure it to the plywood.

3d. For the "legs" of the base, cut two 10" lengths of 2×4 lumber. Position the legs parallel to each other (upright on their long sides) about 7" apart, and place the plywood square on top so it lines up with the outside edges of the legs. Mark and drill pilot holes in the corners of the plywood square and legs, about ¾" in from each outside edge, and use the 4 wood screws to securely attach the base together.

3e. Put the strobe disk back on the motor shaft.

4. MOUNT THE CAMERA (following page)

4a. If your camera has a zoom lens, set it to the widest angle possible. Place your camera on the wooden base so that its lens points at the strobe disk, completely within the slot's path. To keep extra light from entering the lens, the disk and lens should be as close as possible without touching.

4b. Use a pencil to mark the location of the camera's body on the base. Set your camera aside.

4c. Adhere two 3½" strips of velcro tape (hook side) onto the wooden base in the camera body location, parallel to each other and perpendicular to the image plane. The velcro will hold the camera in place and let you make fine adjustments to its position.

4d. Place one strip of velcro (loop side) along the bottom of the camera body. Make sure you can still access the battery compartment.

4e. Position your camera on the base. If it tips forward, slice a round section of cork to place flat under the lens for support;

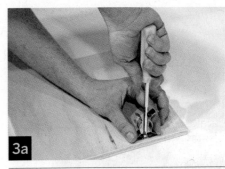

3a

3b

3c

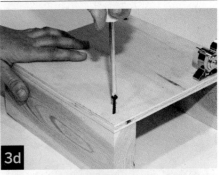

3d

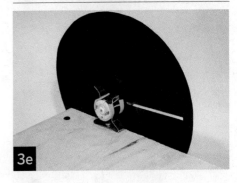

3e

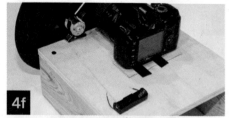

dry-test the fit, then remove the camera and hot-glue the cork to the base.

4f. Put your camera back in position, and locate a place for your battery holder a couple of inches behind your camera and opposite the motor. Remove the battery and use hot glue to attach the battery holder to the base.

4g. Replace the battery and reattach all but one of the alligator test lead connections.

5. MAKE THE HOOD

5a. Find a cardboard box big enough to turn upside down and fit over your whole rig, with room to spare. Use an X-Acto knife to cut out the back wall so you can access the camera. Cut a round opening in the front of the box, to give your camera and strobe disk a clear line of sight.

5b. Place the box over your rig to check the fit. Make sure the camera still has a clear view and that the slot in your strobe disk isn't obscured.

Voilà — your stroboscope is complete!

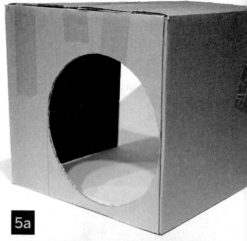

GETTING SET UP

BACKDROP, SUBJECT, AND LIGHTING

Good stroboscope photography requires a black backdrop, preferably fabric, which makes your subject show up clearly without being lost in a bright clutter of background noise. With a black background, light and brightly colored objects will "pop," while dark objects will disappear.

Your subject also needs to be well lit, or else it won't show up. Clamp lights work well and are easy to adjust. With the camera pointing straight toward the backdrop and your subject in between, place 2 clamp lights pointing in from the left and right, respectively, lighting up the subject rather than the backdrop.

You can also set up your stroboscope outside, using sunlight instead of clamp lights. As long as you have a black background and bright light, you're in business.

CAMERA SETTINGS

To set up your camera, temporarily remove the strobe disk and focus your lens on the place where

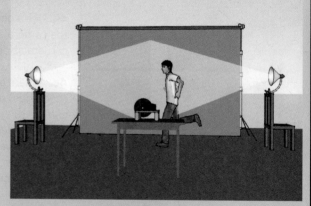

your subject will be moving. Make sure the focus remains set to manual. Here are some good typical settings you can experiment with for starters:

SHUTTER/EXPOSURE: Two seconds. With a too-short exposure, you won't see much happening in the image, but if it's too long, the image will be washed out.

APERTURE: Set this to the lowest number possible to gather the maximum amount of light with each pass of the strobe disk slot.

HAPPY TRAILS!

THE FUN PART

Now you can start to play with strobo-scopic photography. Have a friend press the shutter while you try tossing or bouncing balls, juggling, throwing sticks or paper airplanes, releasing balloons, doing cartwheels, or dancing. Almost anything that moves is fun to photograph with a stroboscope. One of our favorite things to play with is string. Try twirling it in spirals or jumping rope.

Don't be discouraged if your first few images are out of focus or washed out.

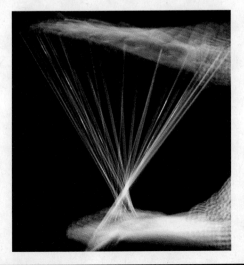

You can solve these issues by adjusting your camera settings, making sure the cardboard hood is in place, and adjusting or adding lights.

The camera will be focused at a single fixed distance, so it helps to mark the floor to remind yourself or your subject to stay in the plane of focus. This will instantly improve your images.

DISK VARIATIONS

Large, slower-moving subjects (like people) look better when the disk only has one slot. However, certain fast-moving subjects — such as thrown objects or vibrating strings — look better if you use a strobe disk with 2 or 3 slots, which doubles or triples the number of exposures per second. Follow Steps 1a–2c to make additional disks that have multiple slots, and make sure that the slots are spaced apart evenly, so the disk stays balanced while it spins. ◪

Nicole Catrett is an exhibit developer at the Exploratorium in San Francisco. You can play with her stroboscope exhibit there.

Walter Kitundu is an artist and bird photographer. You can see more of his work at kitundu.com.

The Diddley Bow

Originally an instrument of the rural South by way of West Africa, the diddley bow is basically a slide guitar stripped down to its most elemental level: one string, no frets, and your ear for tuning. Best of all, you can make it in about 10 minutes. So get playing!

The BeatBearing Tangible Rhythm Sequencer

Music sequencing doesn't get much simpler than Peter Bennett's BeatBearing tangible rhythm sequencer. Move the ball bearings on the grid and you change the beat. The transparent interface highlights which beats are switched on, and what sounds they're playing so you can drop (or maybe gently place) some heavy beats.

SKILL BUILDER:

INDUSTRIAL DESIGN FOR MAKERS

OK, so you've made your project, with an awkward project box, exposed wires and all. Want to bring that lovable mess from the garage into the living room? Or go next-level by manufacturing and selling it? It's time to start thinking like an industrial designer, and Bob Knetzger has some ideas to get you started.

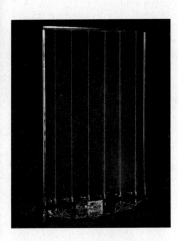

Laser Harp

Laser harps date back to the 80s, but these customizable electronic instruments still ooze futuristic cool. The harp acts as a MIDI controller using an Arduino and off-the-shelf laser pointers to drive an audio synthesizer. Use it to play single notes or activate sound or drum loops. Time to take your prog rock space opera to the next level!

SKILL BUILDER:
LINOCUTS

Similar to woodblock printing, linoleum printing is an easy way to make both simple and complex block prints. It's also versatile and can be used to print on almost any type of paper or fabric. Transfer your design to the linoleum, scrape away the white space, and start printing!

Time-Lapse Movie Setup

Whether it's documenting a busy street corner or blooming flowers, the sped-up world of time-lapse photography is mesmerizing. With an old digital camera and a spare PC, you can continuously capture photos around the clock and compile them into compelling videos.

Bokeh Filter

"Bokeh" comes from the Japanese word for "blur." In photography, the bokeh effect has to do with the aesthetics of out-of-focus areas of the picture. This easy-to-make camera lens attachment allows you to make out-of-focus lights in your pictures appear any shape you want, like hearts, stars, or letters.

The Flame Tube

Forget the fire pit at your next backyard BBQ. This flaming apparatus visualizes sound waves moving through the propane-filled tube, causing the flames to grow larger or smaller. A single tone produces a stationary sine wave, but adding some bass and beats produces a dazzling display of musical pyrotechnics. You know, for science.

Maker SHED™

MAKER'S GUIDE

We're on a mission to help you get started making things. MAKE is all about explaining new technology so that beginners can use it — we literally write the books on this stuff. In the Maker Shed, we curate the best kits, tools, and how-to books that will take you "from zero to maker" in electronics, microcontrollers, embedded computing, 3D printing, and more.

We've done the homework for you. We test and review gear year-round to showcase the latest products that are worth your while. We take the pain out of buying a 3D printer (and our buyer's guide tells you how they really perform). We demystify microcontrollers and single-board computers, offering comparison charts to help you make the choice that's right for you. And we develop great kits based on successful DIY projects from MAKE magazine.

Get it now. Hot products like 3D printers can be hard to source and take forever to ship — ours are ready to roll. And while our Maker Shed stores at Maker Faire are legendary, it's all online, 24/7, at makershed.com.

Contents:

Arduino Esplora

MKSP19

$59

FOR GAMERS AND TINKERERS ALIKE
Find instant gratification in this Arduino disguised as a joystick, with ready-to-use onboard sensors and inputs.

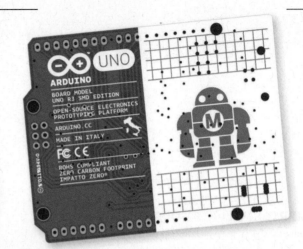

Make: Special Edition Arduino Uno

MKSP99

$29.99

MAKE'S CUSTOM ARDUINO UNO
Order up, special sauce! Get the latest Uno microcontroller (Rev. 3) with our custom MAKE graphics and a great price.

Getting Started with MakerBot

MKBK20

$16.99

BY BRE PETTIS, ANNA KAZIUNAS FRANCE, AND JAY SHERGILL

HexBright Flex Flashlight

MKHB1

$119

THE MAKER'S CHOICE
A programmable flashlight that has a GitHub repository to share hacks, like varying brightness based on angle held.

Arduino GSM Shield

MKSP21

$99

CONNECT TO THE GSM CELLULAR NETWORK
Use this shield to handle phone calls and SMS. We can't wait to see the cool projects that come out of this technology!

Getting Started with Raspberry Pi (Bundled with the Board)

MKRPI2

$54.99

BY MATT RICHARDSON AND SHAWN WALLACE
You'll be up and running quickly with this board-and-book combo pack.

Maker SHED **MAKER'S GUIDE**

3D PRINTING

Afinia H-Series

DSAF1

$1,599

EASY TO USE, WITH FEATURE-RICH SOFTWARE AND IMPRESSIVE PRINT QUALITY

This accurate and reliable printer arrives fully assembled and only takes a few minutes to set up, making it ideal for beginners. Documentation is straightforward and easy to follow, with an in-depth manual and a one-page Quick Start Guide for those wanting to jump right in. Sturdy enough for transport, it weighs just 11lbs.

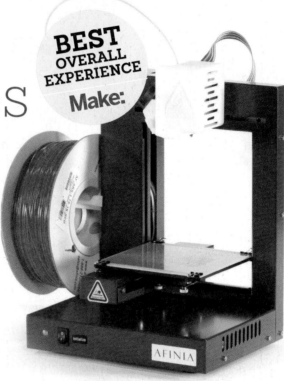

BEST OVERALL EXPERIENCE Make:

Make: Ultimate Guide to 3D Printing

MKBK4

$9.99

CHOOSE THE 3D PRINTER THAT'S RIGHT FOR YOU

The in-depth DIY guide.

BEST IN CLASS: PREMIUM Make:

MakerBot Replicator 2

DSMB03

$2,199

FAST, DEPENDABLE, AND VERY QUIET WITH A CLEAN, INTUITIVE USER INTERFACE

This sleek-looking machine is easy to set up and produces consistently attractive prints. Its responsive LCD panel allows for easy control and monitoring, and a Cold Pause feature halts the print, cools the extruder, and waits for you to resume where you left off, which could come in handy in more than a few situations.

MakerBot Replicator 2X

MKMB04

$2,799

MAKERBOT'S NEWEST MODEL, WITH DUAL EXTRUDERS TO OPEN UP EVEN MORE POSSIBILITIES

This impressive machine has a heated bed for printing with ABS, as well as an acrylic enclosure to keep the heat in and little fingers out. With a dual extruder, you can print objects with two different colors or two different materials.

BEST VALUE
Make:

Printrbot Jr.

DSPB1
$499

SIMPLE, SMALL, AND PORTABLE, COMES ASSEMBLED AT A PRICE WITHIN REACH OF CASUAL USERS

This easy-to-set-up, easy-to-use printer is small enough to fold up and slip into a backpack. Unique to this machine: the option of using a rechargeable lithium polymer battery when printing in the field.

MAKER SHED'S **BEST SELLER**

BEST IN CLASS: ENTRY LEVEL **Make:**

Printrbot LC V2 Kit

DSPB2
$649

FAST AND INEXPENSIVE, WITH THE ABILITY TO TWEAK AND UPGRADE

This compact, flexible, build-it-yourself printer's open frame design makes it extendable, allowing you to modify it to create larger prints.

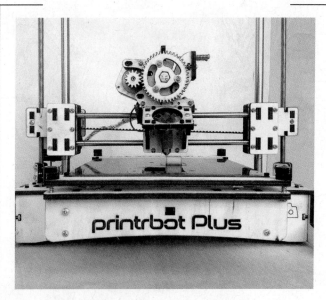

Printrbot Plus V2 Kit

DSPB3
$799

SOLID AND AFFORDABLE, WITH A LARGE BUILD AREA AND ROOM TO GROW

This kit's straightforward assembly makes it a great weekend project, with plenty of customizable options for the relatively tech-savvy maker.

3D PRINTING

BEST
DOCUMENTATION
Make:

3D Systems
Cube

DS3DS01
$1,299

STYLISH, WITH A WELL LAID-OUT TOUCHSCREEN CONTROL PANEL AND WI-FI PRINTING

This reliable, easy-to-use printer is well documented and looks great on your desk. The included "magic glue" helps keep prints stuck to the build surface, and washes away with water to cleanly release them. Users can configure nozzle height, Wi-Fi settings, and other details using the touchscreen interface, which also shows the status of current print progress and other system indicators.

BEST
IN CLASS:
MIDRANGE
Make:

Type A Series 1

MKTA1*
$1,400

ACCURATE, FAST, AND RELIABLE, WITH A HUGE BUILD AREA AND LOTS OF POTENTIAL

This printer is designed for speed and accuracy and works great right out of the box. You can also download the pertinent case and equipment files to build your own and make modifications.

*While supplies last.

BEST
OPEN
HARDWARE
Make:

Ultimaker 3D Printer Kit

DSUM1*
$1,599

FAST, ACCURATE, AND MODIFIABLE, WITH STRAIGHTFORWARD BUILD INSTRUCTIONS

This high-resolution printer with a large build platform is designed for speed. Plus, users can hack it, making it a perfect machine for the tinkering maker.

*While supplies last.

REPLICATOR 2/2X

$2,199/$2,799

Print Volume
11.2"×6"×6.1"
Print Speed (per Mfr.)
80–100mm/sec
Print Material
PLA (2) / ABS, PLA (2X)
OS Supported
Windows, Mac, Linux
Print without computer? SD card
makerbot.com

ULTIMAKER

$1,599 KIT

Print Volume
8.3"×8.3"×8.3"
Print Speed (per Mfr.)
150mm/sec
Print Material
PLA
OS Supported
Windows, Mac, Linux
Print without computer? Optional, SD card
ultimaker.com

AFINIA H-SERIES

$1,599

Print Volume
5.5"×5.5"×5.3"
Print Speed (per Mfr.)
3–30mm³/sec
Print Material
ABS, PLA
OS Supported
Windows, Mac
Print without computer? Onboard storage
afinia.com

TYPE A SERIES 1

$1,400

Print Volume
9"×9"×9"
Print Speed (per Mfr.)
90mm/sec
Print Material
ABS, PLA, PVA
OS Supported
Windows, Mac, Linux
Print without computer? No
typeamachines.com

CUBE

$1,299

Print Volume
5½"×5½"×5½"
Print Speed (per Mfr.)
15mm³/sec
Print Material
ABS, PLA
OS Supported
Windows
Print without computer? Wi-Fi, USB stick
cubify.com/cube

PRINTRBOT PLUS

$799 KIT

Print Volume
8"×8"×8"
Print Speed (per Mfr.)
70mm/sec
Print Material
ABS, PLA
OS Supported
Windows, Mac, Linux
Print without computer? SD card
printrbot.com

PRINTRBOT LC

$649 KIT

Print Volume
6"×6"×6"
Print Speed (per Mfr.)
70mm/sec
Print Material
ABS, PLA
OS Supported
Windows, Mac, Linux
Print without computer? SD card
printrbot.com

PRINTRBOT JR.

$499

Print Volume
4"×4"×4"
Print Speed (per Mfr.)
70mm/sec
Print Material
PLA
OS Supported
Windows, Mac, Linux
Print without computer? SD card
printrbot.com

ARDUINO BOARDS

Make:
Ultimate Arduino Microcontroller Project Pack
MSUMP1
$119.99

A SMART PURCHASE FOR ANYONE SERIOUS ABOUT LEARNING MICROCONTROLLERS
More than 100 electronics components and parts are packaged along with an Arduino, giving you the means to create a multitude of projects right out of the box.

Make: Getting Started with Arduino Kit v3.0

MSGSA
$64.99

LEARN HOW TO USE THE ARDUINO MICROCONTROLLER
Write simple programs — such as blinking LEDs and reading sensor inputs — and advance to creating custom code to control motors, lights, and other outputs.

Getting Started with Arduino 2nd Edition

BY MASSIMO BANZI
A perfect companion to the kit.

MBK1
$14.99

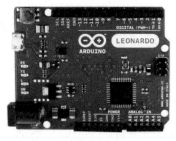

Arduino Leonardo with Soldered Headers
MKSP15
$24.99

OPENS DOORS TO PRODUCT DEVELOPMENT
This board has HID support, which allows it to appear to any connected computer as a mouse and keyboard.

Arduino Due
MKSP16
$49.99

THE FIRST ARM PROCESSOR-POWERED ARDUINO
This powerful board packs many new features in a Mega-sized form factor.

Arduino Mega 2560
MKSP5
$65

FOR MORE INVOLVED PROJECTS This board offers more inputs/outputs and 4x the memory of the standard Uno.

Arduino Uno

MKSP11

$34.⁹⁹

IN ITS THIRD REVISION, THIS IS THE CURRENT "OFFICIAL" ARDUINO BOARD
With this microcontroller, you can create interactive objects that can sense inputs from switches, sensors, and computers — and then control motors, lights, and other physical outputs in the real world. Features driverless USB-to-serial and auto power switching.

PICKING YOUR FIRST MICROCONTROLLER
Choose the right controller for your project and your skill level. *By Tom Igoe*

FOCUS ON EASE OF USE. You're learning to program and learning to build a computer. Starting simple will keep your enthusiasm high.

DON'T BE SEDUCED BY FEATURES. Apollo spacecraft made it to the moon and back with less processing power than most microcontrollers have. Don't be tempted by the fastest, or the one with the most memory or I/O, at the expense of simplicity.

WHAT'S THE "GETTING STARTED" GUIDE LIKE? The time you're most likely to give up on microcontrollers is in the first hour. The Getting Started guide should take you from zero to blinking an LED or reading a switch in a short time. Read it before you buy. If you don't understand any of it, or it doesn't exist, be wary.

HOW COMPLICATED IS THE PROGRAMMING ENVIRONMENT? The program in which you write your code, called the integrated development environment (IDE), should be easy to understand. Download the IDE and

check it out before you buy hardware. If you don't feel you can understand it quickly, keep looking around.

HOW EXPANDABLE IS THE PROGRAMMING ENVIRONMENT? I/O boards and very simple languages make getting started easy, but you'll reach their limits quickly. If you're experienced in programming or electronics, you want to feel liberated by a platform's simplicity, not limited by it.

IS THE PROGRAMMING ENVIRONMENT FREE? If not, don't bother. There are too many good free environments for a beginner to bother paying for the software alone.

WHAT'S THE COMMUNITY KNOWLEDGE BASE LIKE? You're not just getting hardware, you're getting a community. Every controller has websites and email lists dedicated to its use; check them out, look at the code samples and application notes, read a few discussion threads. Do a few web searches for the microcontroller you're considering. Is there a lot of collected

knowledge available in a form you understand? If nobody besides you is using your controller, you'll find it much harder to learn, no matter how great its features are.

HOW EASY IS IT TO ADD EXTRA COMPONENTS? If there's a particular component you want to work with, check to see if someone's written an example for how to use it with the controller you're considering. Most controllers offer 16 or so I/O connections, which is plenty enough to get started. Tools for expanding your I/O, such as shift registers and multiplexers, are compatible with most controllers.

IS YOUR OPERATING SYSTEM SUPPORTED? Most microcontroller manufacturers focus on the Windows operating system. Some have third-party support for Mac OS X and Linux. Learning from friends is common, and being able to have the same user experience on different operating systems is helpful for that.

RASPBERRY PI & MORE

Raspberry Pi Starter Kit
Includes Raspberry Pi!

MSRPIK

$129.⁹⁹

POWER YOUR NEXT PROJECT WITH THIS TINY, HACKABLE COMPUTER

This credit card-sized computer runs Linux and has many of the capabilities of a traditional PC. Our Starter Kit includes a copy of the *Getting Started with Raspberry Pi* book, as well as all the proper peripherals and add-ons to get you up and running in all sorts of maker applications. You'll still need to add your own USB keyboard and mouse, but otherwise, this kit includes all the components you need to get started.

AlaMode for Raspberry Pi
MKWY1

$49.⁹⁹

GIVE YOUR PI THE EXPANSION AND EASE OF USE OF AN ARDUINO

This Arduino-compatible board connects to the Pi, giving you the freedom to write programs to control or monitor your Arduino application in any language you like.

BeagleBone Black
MKCCE3

$45

THE CONNECTIVITY OF ARDUINO, WITH THE PROGRAMMABILITY OF RASPBERRY PI

Ready to use right out of the box, this small, low-cost, high-expansion standalone Linux computer is perfect for hackers and tinkerers who need a little more "oomph" to run complex electronics projects.

MICROCONTROLLER COMPARISON CHART

BOARD	ARDUINO UNO	ARDUINO LEONARDO	ARDUINO DUE	MINTDUINO	RASPBERRY PI	BEAGLEBONE BLACK
PRICE	$34.99	$24.99	$49.99	$24.99	$39.99 (N/A)	$45
STARTER KIT	$64.99				$129.99	
QUICK SUMMARY	Current "official" Arduino USB board, driverless USB-to-serial, auto power switching	Somewhat experimental Arduino with HID support for mouse or keyboard emulation	Newest Arduino based on a powerful ARM processor. Packs many new features in a Mega-sized form factor	An Arduino-compatible board you build yourself on a breadboard	Single-board Linux computer with video processing and GPIO ports	Next-gen, ARM-based, hardware hacker-focused Linux board; RaspPi programmability + Arduino connectivity
SPECIAL FEATURES	Onboard USB controller	HID emulation, USB, SPI on ISP header	Android ADK Support, 2 12bit ADC / DAC, USB host, CAN BUS support	DIY Arduino!	HD-capable video processor, HDMI and composite outputs, onboard Ethernet	Onboard USB host and Ethernet, 2GB onboard storage, HDMI output
PROCESSOR	ATmega328	ATmega32U4	32-bit SAM3X8E ARM Cortex-M3	ATmega328	ARM1176JZF-S	Sitara AM3359A ARM Cortex-A8
PROCESSOR SPEED	16MHz	16MHz	84MHz	16MHz	700MHz	1GHz
ANALOG PINS	6	12	12	6	None (no onboard ADC)	7
DIGITAL PINS	14 (6 PWM)	20 (7 PWM)	54 (12 PWM)	14 (6 PWM)	8 digital GPIO	65 GPIO (8 PWM)
MEMORY	SRAM 2KB, EEPROM 1KB	SRAM 2.5 KB, EEPROM 1 KB	SRAM 96 KB	SRAM 2KB, EEPROM 1KB	RAM 512MB	DRAM 512MB DDR3L, eMMC 2GB
PROGRAMMING LANGUAGE	Arduino / C variant	Arduino / C variant	Arduino / C variant	Arduino / C variant	Any language supported by a compatible Linux distribution (such as Raspbian or Occidentalis)*	Any language supported by a compatible Linux distribution (Ångström, Ubuntu, Android, etc.).* Pre-loaded with Ångström
PROGRAMMER	USB, ISP	USB, ISP	USB, ISP	Requires programmer like FTDI Friend	Runs any of the Linux-compatible text editors and IDEs on the board	Runs any of the Linux-compatible text editors and IDEs; supports Web IDE and BoneScript
EXPANSION	Shield compatible	Most shields	Some shields (3V only)	N/A	Breakout boards such as Pi Plate, Pi Cobbler	Capes
ASSEMBLED OR KIT	Assembled	Assembled	Assembled	Kit	Assembled	Assembled

*Including Python, Scratch, Perl, Java, JavaScript/Node, C, C++, and Ruby.

BEGINNER ROBOTS

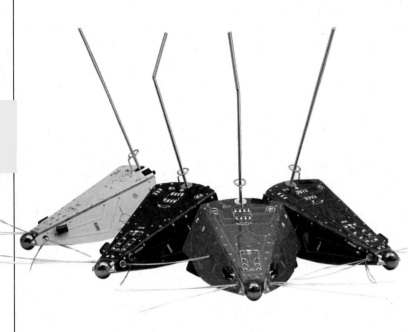

Mousebot Kit

MKSB001

$39.95

A LIGHT-CHASING ROBOT THAT'S EASY TO SOLDER AND PERFECT FOR BEGINNERS

Herbie the Mousebot is a favorite here at MAKE and has been a staple in the Shed for years. This full-fledged robot can explore its environment, and even chase around other Herbies. He has an ingenious body made from three electrically connected printed circuit boards, and is a graceful step up from basic skills to slightly more challenging soldering.

Big Bad Beetlebot Kit

MKSB019

$39.95

THIS OBSTACLE-AVOIDING ROBOT IS A GREAT PARENT-CHILD PROJECT

Made of simple components, this autonomous robot can sense its environment and navigate a space. Antenna-equipped switches cross-wired to two DC motors power this minimalist bug, which has been featured as a popular MAKE magazine and Instructables project. No soldering required.

Make: Rovera Arduino Robot Kit

← **2WD**
MSROB2W
$169.⁹⁹

4WD
MSROB4W
$194.⁹⁹

DELVE INTO THE WORLDS OF BOTH MICROCONTROLLERS AND ROBOTICS

Make a robot that can serve as the foundation for all sorts of robotic experimentation. Available in 2WD and 4WD models, both kits come with the book *Make an Arduino-Controlled Robot*, which will walk you through building, programming, and expanding each one.

Make an Arduino-Controlled Robot
MKBK3
$19.⁹⁹

BY MICHAEL MARGOLIS

EZ-Robot Complete Kit

MKEZ2
$169

A DREAM-COME-TRUE SET OF PARTS FOR HOBBY ROBOTICISTS

This kit lets you turn just about anything (R/C vehicles, motorized "dumb" robot toys, little sister's doll collection) into a robot. And the downloadable (Windows-only) software interface makes it easy to program and control your bots. Combine this with one of our robot platform kits, and you're well on your way to creating the race of intelligent machines that are going to rise up one day and take over the world.

MAKE ELECTRONICS

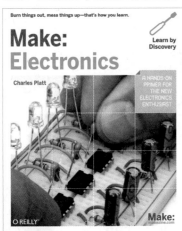

Make: Electronics Book

MKBK2

$34.⁹⁹

BURN THINGS OUT, MESS THINGS UP — THAT'S HOW YOU LEARN

Complete the 36 experiments in this clearly written book, and you'll have a solid knowledge of electronics and the confidence to venture into complex projects.

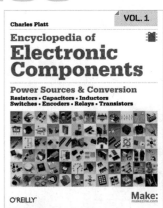

Encyclopedia of Electronic Components Vol. 1

MKBK17

$24.⁹⁹

THIS REFERENCE PUTS RELIABLE, KEY INFORMATION RIGHT AT YOUR FINGERTIPS

Packed with photographs, schematics, and diagrams, this well-organized book explains each component, how it works, why it's useful, and what variants exist.

Make: Electronics Components Pack 1

MECP1

$109.⁹⁹

LEARN VOLTAGE, AMPERAGE, AND RESISTANCE THROUGH HANDS-ON EXPERIMENTS

This 250-piece companion pack gives you all the components you need to complete experiments from the first two sections of our popular book, *Make: Electronics*.

Make: Electronics Components Pack 2

MECP2

$129.⁹⁹

EXPERIENCE THE CIRCUIT FIRST, THEN LEARN THE THEORIES BEHIND IT

This more than 130-piece companion pack gives you the rest of the components you need to complete experiments 12–24 in our popular book, *Make: Electronics*.

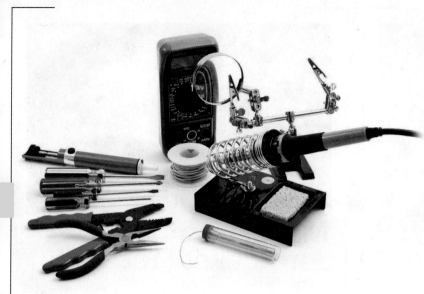

Make:
Electronics Deluxe Toolkit

MKEE1
$129.99

A COMPLETE SET OF TOOLS TO GET STARTED WITH ELECTRONICS AND SOLDERING

Great by itself, or as a companion to the *Make: Electronics* book and Components Packs, this kit contains more than 20 tools in a handy case.

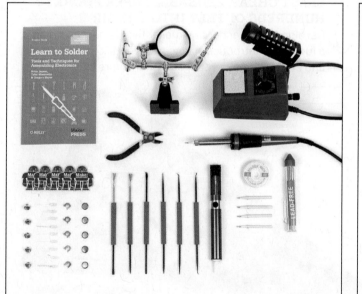

Make: Soldering Kit

MSGSWS
$64.99

GO FROM NOVICE TO PROUD BLINKY BADGE WEARER IN AN AFTERNOON

With this collection of tools and our *Learn to Solder* book, you'll be mastering this skill in no time. Added bonus: we've included five of our Learn to Solder badges.

Mintronics:
Survival Pack

MSTIN2
$24.99

BECAUSE YOU NEVER KNOW WHEN YOU MIGHT NEED TO MACGYVER SOMETHING

This jam-packed mint tin contains more than 60 useful components for making, hacking, and modifying electronic circuits and repairs on the go.

KITS

Elev-8 Quadcopter Kit

MKPX23

$599.99

**FLYING ROBOTIC PLATFORM IS LIFTED AND
PROPELLED BY FOUR FIXED ROTORS**

Large enough for outdoor flight, with room for payload
and attachments, this kit is an inexpensive way to get
into the quadcopter arena.

Make: Compressed Air Rockets Kit

MKRS1

$69.99

**BLAST CHEAP, REUSABLE PAPER ROCKETS
HUNDREDS OF FEET INTO THE AIR**

Developed by teacher and maker Rick Schertle, this kit is
a fun way to introduce science concepts, basic soldering,
and electronics to a youngster or an entire classroom.

Calculator Kit

MKSKL16

$44.95

CONSTRUCT YOUR OWN OLD-SCHOOL TOOL

This quick-to-assemble DIY calculator kit is pre-pro-
grammed so it will work as soon as you're done soldering
and put in the battery.

Brooklyn Aerodrome Flying Wing Kit

MSFW1

$299

**CRASHING IS PART OF THE FUN WITH THIS
DURABLE R/C AIRPLANE**

Designed to be maneuverable and easily repairable,
this kit plane is optimized for gusty winds, small flying
spaces, and rough landing spots.

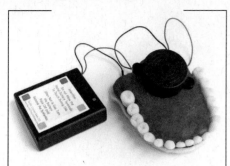

Squishy Circuits Kit

MKSC1

$24.99

EXPLORE ELECTRONICS USING A FUN, FAMILIAR FORM — PLAY DOUGH

Light LEDs, make noise, and run motors by simply plugging them into conductive dough.

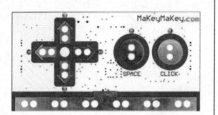

Makey Makey

MKJL1

$49.99

THIS ARDUINO-BASED DEVICE TURNS NEARLY ANYTHING INTO A COMPUTER KEY

Just attach the included alligator clips to food, people, liquids, or any other conductive material for a new level of interactivity. Make a drum kit from oranges, a keyboard from bananas, and jam out.

Bare Conductive House Kit

MKBC4

$19.99

CREATE PAPER HOUSES THAT LIGHT UP IN THE DARK

Create two light-sensitive paper houses using electrically conductive paint.

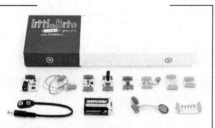

LittleBits Starter Kit v0.3

MKLB2

$89

SMALL MAGNETS MAKE PROTOTYPING SOPHISTICATED ELECTRONICS A SNAP

Each bit has a simple, unique function (light, sound, sensors, buttons, thresholds, pulse, motors, etc.) and modules connect to make larger circuits.

Make: BrushBot Party Pack

MSBBRP

$24.99

TURN A TOOTHBRUSH INTO A ROBOT

Easy to make, fun to personalize, and good, goofy fun to race, these vibrating robots are great for groups, parties, or schools. Kit makes 12 BrushBots.

Make: SpinBot Kit

MSRSPIN

$24.99

BUILD A SIMPLE, EASY-TO-ASSEMBLE VIBRATING ARTBOT

This triple-armed, pen- or chalk-grasping bot spins in circles to draw elaborate geometric shapes. Just add markers and a battery.

FUN TOOLS

Make: Circuit Breaker Leatherman

MKLTM3

$36

THIS MINI WIRE STRIPPER TOOL FITS ON YOUR KEY CHAIN

The Circuit Breaker is the perfect companion for electronics, mobile fixing, hacking, and MacGyvering. This Squirt model also features built-in scissors.

54-Piece Bit Driver Kit

MKIF2

$24.95

GREAT FOR ACCESSING LAPTOPS AND OTHER SMALL ELECTRONICS

This kit includes all the current common and specialty bits, and features a 60mm extension as well as a 130mm flexible extension for those hard-to-reach areas.

Jackknife Pocket Lock Pick Set

MKSD02

$39.99

A PERFECT SIZE TO KEEP AT THE READY

This high-quality set features tempered stainless steel picks, knurled stainless steel set screw for a positive lockup, and a hard alloy handle for increased durablity.

TOOOL Emergency Lock-Pick Card

MKLPO2

$29.99

AVOID BEING LOCKED OUT WITH THIS SNAP-APART SET

Fits in your wallet, handy when a situation arises, and once removed, you can place the lock picks on your key chain.

Beginner's Lock-Picking Blend Set

MKLPO1

$39.99

LEARN THE ART OF PICKING LOCKS WITH THIS 8-TOOL SET

Hand selected by The Open Organisation Of Lockpickers, these tools open the majority of pin tumbler locks in use today.

Locksmith School in a Box

MKSD01

$99.99

MASTER PIN-TUMBLER LOCKS WITH THIS PROGRESSIVE LEARNING SYSTEM

Featuring five increasingly difficult lock cylinders, this kit includes four picks, one tension tool, and an instructional book.

WHAT YOU NEED TO GET STARTED IN HOBBY ELECTRONICS.

By Charles Platt

Charles Platt is the author of *Make: Electronics*, an introductory guide for all ages. He is completing a sequel, *Make: More Electronics*, and is also the author of the *Encyclopedia of Electronic Components, Volume 1*. Volumes 2 and 3 are in preparation.

The Basics

First, you'll need a breadboard. You can call it a "prototyping board," but this is like calling a battery a "power cell." Search online for "breadboard" and you'll find more than a dozen products, all of them for electronics hobbyists, and none of them useful for doing anything with bread.

A breadboard is a plastic strip perforated with holes 1⁄10" apart, which happens to be the same spacing as the legs on old-style silicon chips — the kind that were endemic in computers before the era of surface-mounted chips with legs so close together only a robot could love them. Fortunately for hobbyists, old-style chips are still in plentiful supply and are simple to play with.

Your breadboard makes this easy. Behind its holes are copper conductors, arrayed in hidden rows and columns. When you push the wires of components into the holes, the wires engage with the conductors, and the conductors link the components together, with no solder required.

Figure A (following page) shows a basic breadboard. You insert chips so that their legs straddle the central groove, and you add other components on either side. You'll also want to buy a matching printed circuit board (PCB) that has the same pattern of copper connectors as the breadboard. First use the breadboard to make sure everything works, then transpose the parts to the PCB, pushing their wires through from the top. You immortalize your circuit by soldering the wires to the copper strips.

Soldering, of course, is the tricky part. As always, it pays to get the right tool for the job. I never used to believe this, because I grew up in England, where "making do with less" is somehow seen as a virtue. When I finally bought a 15-watt pencil-sized soldering iron with a very fine tip (**Figure C**), I realized I had spent years punishing myself. You need that fine-tipped soldering iron, and thin solder to go with it. You also need a loupe (**Figure D**), a little magnifier. A cheap plastic one is sufficient. You'll use it to make sure the solder you apply to the PCB hasn't run across any of the spaces separating adjacent copper strips, creating short circuits.

Short circuits are the #2 cause of frustration when a project that worked perfectly on a breadboard becomes totally uncommunicative on a PCB. The #1 cause of frustration (in my experience, anyway) would be dry joints.

Any soldering guide will tell you to hold two metal parts together while simultaneously applying solder and the tip of the soldering iron. If you can manage this far-fetched anatomical feat, you must also watch the solder with supernatural close-up vision. You want the solder to run like a tiny stream that clings to the metal, instead of forming beads that sit on top of the metal. At that precise moment, you remove the soldering iron. The solder solidifies, and the joint is complete.

You get a dry joint if the solder isn't quite hot enough. Its crystalline structure lacks integrity and crumbles under stress. If you've joined two wires, it's easy to test for a dry joint: you can pull them apart easily. On a PCB, it's another matter. You can't test a chip by trying to pull it off the board, because the good joints on most of its legs will compensate for any bad joints.

You'll use your loupe to check for the bad joints. You may see a wire-end perfectly centered in a PCB hole, with solder on the wire, solder around the hole, but no solder actually connecting the two. This gap of 1⁄100" is enough to stop everything from working, but you'll need a good desk lamp and magnification to see it.

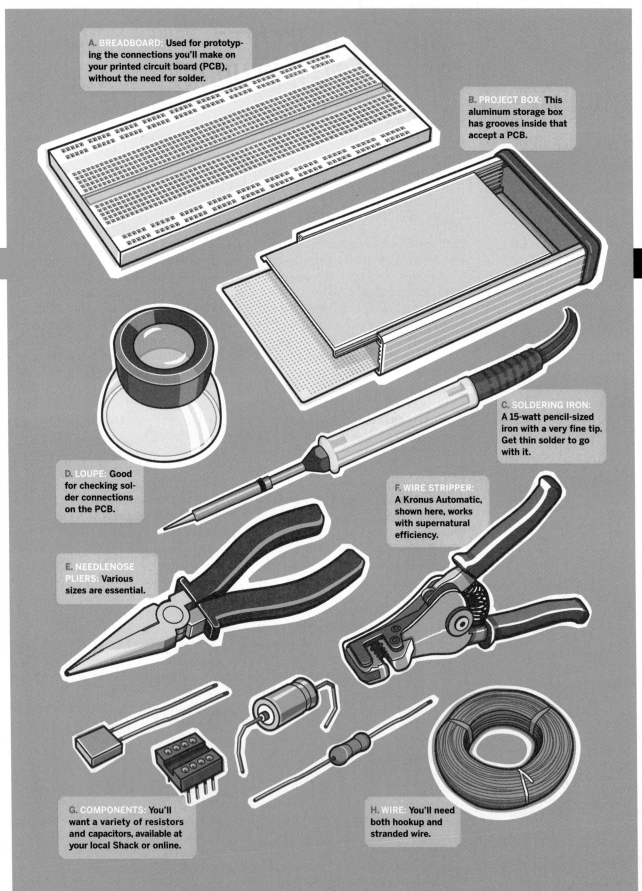

A. BREADBOARD: Used for prototyping the connections you'll make on your printed circuit board (PCB), without the need for solder.

B. PROJECT BOX: This aluminum storage box has grooves inside that accept a PCB.

C. SOLDERING IRON: A 15-watt pencil-sized iron with a very fine tip. Get thin solder to go with it.

D. LOUPE: Good for checking solder connections on the PCB.

F. WIRE STRIPPER: A Kronus Automatic, shown here, works with supernatural efficiency.

E. NEEDLENOSE PLIERS: Various sizes are essential.

G. COMPONENTS: You'll want a variety of resistors and capacitors, available at your local Shack or online.

H. WIRE: You'll need both hookup and stranded wire.

Illustration by Damien Scogin

A Few Components and Tools

Just as a kitchen should contain eggs and orange juice, you'll want a variety of resistors and capacitors (**Figure G**). Your neighborhood Radio Shack can sell you prepackaged assortments, or you can shop online at makershed.com.

After you buy the components, you'll need to sort and label them. Some may be marked only with colored bands to indicate their values. With a multimeter (a good one costs maybe $50) you can test the values instead of trying to remember the color-coding system. For storage I like the kind of little plastic boxes that craft stores sell to store beads.

For your breadboard you'll need hookup wire (**Figure H**). This is available in precut lengths, with insulation already stripped to expose the ends. You'll also need stranded wire to make flexible connections from the PCB to panel-mounted components such as LEDs or switches. To strip the ends of the wire, nothing beats the Kronus Automatic Wire Stripper (**Figure F**), which looks like a monster but works with supernatural efficiency, letting you do the job with just one hand.

Needlenose pliers (**Figure E**) and side cutters of various sizes are essential, with perhaps tweezers, a miniature vise to hold your work, alligator clips, and that wonderfully mysterious stuff, heat-shrink tubing (you'll never use electrical tape again).

If this sounds like a serious investment, it isn't. A basic workbench should entail no more than a $250 expenditure for tools and parts. Electronics is a much cheaper hobby than more venerable crafts such as woodworking, and since all the components are small, it consumes very little space.

For completed projects you'll need project boxes (**Figure B**). You can settle for simple plastic containers with lids, but I prefer something a little fancier. Hammond Instruments makes a brushed aluminum box with a lid that slides out for access and grooves inside the box for a PCB. My preferred box has a pattern of conductors emulating three breadboards put together. This is big enough for ambitious projects involving multiple chips.

Learn the Rules

Read a basic electronics guide, like my *Make: Electronics* (makershed.com), to learn the relationships between ohms (Ω), amperes (A), volts (V), and watts (W), so that you can do the numbers and avoid burning out a resistor with excessive current or an LED with too much voltage. And follow the rules of troubleshooting:

» LOOK FOR DEAD ZONES

This is easy on a breadboard, where you can include extra LEDs to give a visual indication of whether each section is dead or alive. You can use piezo beepers for this purpose, too. And, of course, you can clip the black wire of your meter to the negative source in your circuit, then touch the red probe (carefully, without shorting anything out!) to points of interest.

If you get an intermittent reading when you flex the circuit board gently, almost certainly you have a dry joint somewhere, making and breaking contact. More than once I've found that a circuit that works fine on a naked PCB stops working when I mount it in a plastic box, because the process of screwing the board into place has flexed it just enough to break a connection.

» CHECK FOR SHORT CIRCUITS

If there's a short, current will prefer to flow through it, and other parts of the circuit will be deprived. They'll show much less voltage than they should.

Alternatively you can set your meter to measure amperes and then connect the meter between one side of your power source and the input point on your circuit. A zero reading on the meter may mean that you just blew its internal fuse because a short circuit tried to draw too much current.

» CHECK FOR HEAT-DAMAGED COMPONENTS

This is harder, and it's better to avoid damaging the components in the first place. If you use sockets for your chips, solder the empty socket to the PCB, then plug the chip in after everything cools. When soldering delicate diodes (including LEDs), apply an alligator clip between the soldering iron and the component. The clip absorbs the heat.

Tracing faults in circuits is truly an annoying process. On the upside, when you do manage to put together an array of components that works properly, it usually keeps on working cooperatively, without change or complaint, for decades — unlike automobiles, lawn mowers, power tools, or, for that matter, people.

To me this is the irresistible aspect of hobby electronics. You end up with something that's more than the sum of its parts — and the magic endures.

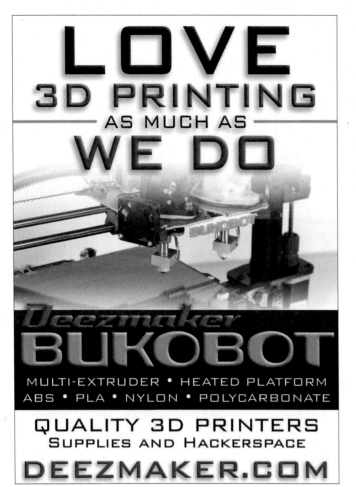
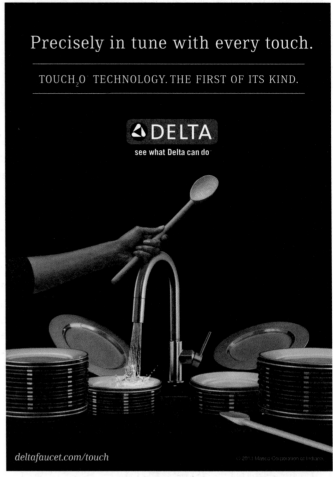

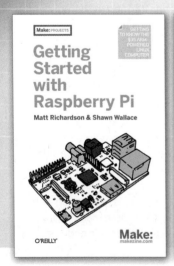

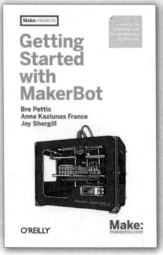
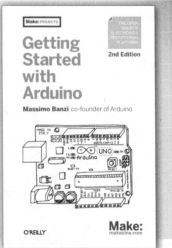

HOW KALEIDOSCOPE TOON

EIDOS FORM

KALOS = BEAUTY

SKOPEO = TO LOOK

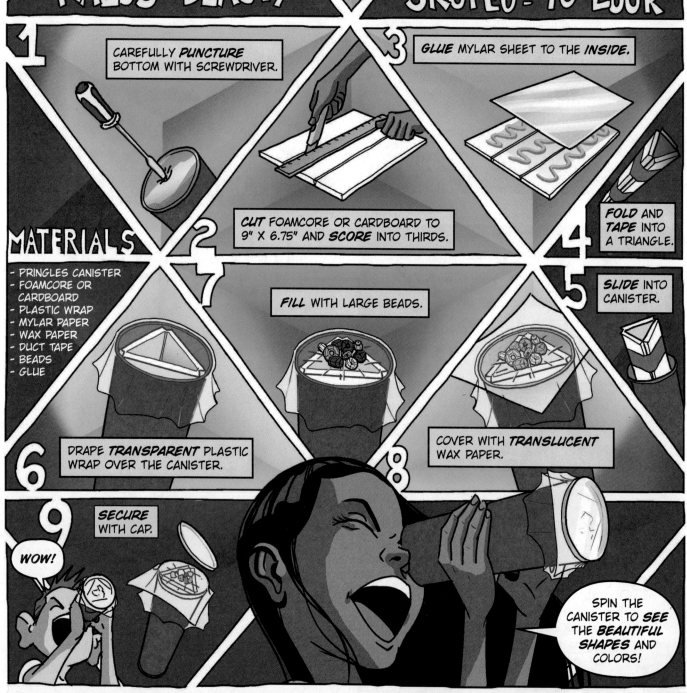

1 CAREFULLY *PUNCTURE* BOTTOM WITH SCREWDRIVER.

2 *CUT* FOAMCORE OR CARDBOARD TO 9" X 6.75" AND *SCORE* INTO THIRDS.

3 *GLUE* MYLAR SHEET TO THE *INSIDE*.

4 *FOLD* AND *TAPE* INTO A TRIANGLE.

5 *SLIDE* INTO CANISTER.

6 DRAPE *TRANSPARENT* PLASTIC WRAP OVER THE CANISTER.

7 *FILL* WITH LARGE BEADS.

8 COVER WITH *TRANSLUCENT* WAX PAPER.

9 *SECURE* WITH CAP.

MATERIALS
- PRINGLES CANISTER
- FOAMCORE OR CARDBOARD
- PLASTIC WRAP
- MYLAR PAPER
- WAX PAPER
- DUCT TAPE
- BEADS
- GLUE

WOW!

SPIN THE CANISTER TO *SEE* THE *BEAUTIFUL SHAPES* AND COLORS!

Training Camp for Makers!

STARTS JUNE 1ST SIGN UP TODAY!

Our virtual "boot camp" for makers.

Online courses include skill-building, tips, and tricks and how-to tutorials in the most popular areas. Classes are multi-session, culminating in a project at the end of the course. **The best part: it's taught by makers for makers.**

FIRST COURSES:
Introduction to Arduino
Introduction to Raspberry Pi

Make:
makezine.com/trainingcamp

VICTORIAN TOYS AND FLATLAND ROCKETS

Invented & drawn by Bob Knetzger

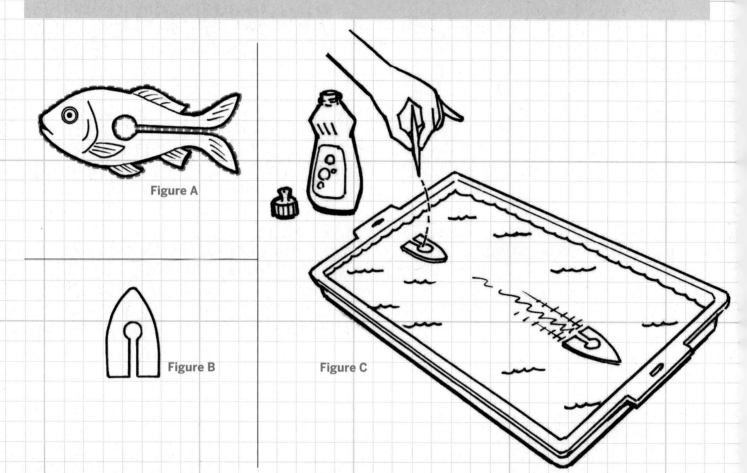

Figure A

Figure B

Figure C

HERE'S AN UPDATE OF A VICTORIAN PLAYTHING. Cut out the fish on the dotted line and float it on a pan of water. Place a single drop of olive oil in the circle. The oil quickly spreads out the slit and across the water. The fish "swims" in the opposite direction, like an exhaust-spewing rocket subject to Newton's third law of motion. Sadly, the soggy paper fish is only good for just a single use (**Figure A**).

NOW TRY THIS NEW, MORE DURABLE VERSION: find a flexible lid from a margarine or yogurt container. Look for the recycling symbol 2 or 4 for low- or high-density polyethylene. (PE is one of the few plastics that floats.) Use a paper punch to make a small circular hole, then cut out the "rocket" shape, same size as shown in **Figure B**.

Float the rocket in a pan of clean water. Dip the tip of a toothpick in detergent and momentarily touch it inside the rocket's round hole. As the detergent dissolves, it spreads down the slit and out along the surface of the water — the rocket shoots forward! Touch it again. After a time or two, you'll have to change the water for the effect to work again. Visit makezine.com/maker-projects to see a video demo.

Another force is also at work: the Marangoni effect, the difference in surface tensions created by the molecules of detergent as they make the water slipperier and "wetter." The surface tension is reduced behind the rocket, causing the water in front to contract, pulling the rocket forward (**Figure C**).

These tensions, forces, and actions all exist at the single-molecule-thick surface of the water — similar to the two-dimensional world in Edwin Abbott's Victorian-era book, *Flatland: A Romance of Many Dimensions.*